Assay Development and Evaluation

A Manufacturer's Perspective

Jan S. Krouwer, PhD

Krouwer Consulting

2101 L Street, NW, Suite 202
Washington, DC 20037-1558

1 2 3 4 5 6 7 8 9 0   VGI   03 02 01

Printed in the United States of America

Library of Congress Cataloging-in-Publication Data

Krouwer, Jan S.
   Assay development and evaluation : a manufacturer's perspective / Jan S. Krouwer.
     p. ; cm.
   Includes index.
   ISBN 1-890883-81-6 (alk. paper)
    1. Diagnostic reagents and test kits. 2. Diagnostic reagents industry. I. Title.
  RB125.K76 2002
  615'.19–dc21

                                       2002074695

Contents

Preface

At Technicon Instruments Corp., the leader in the field of diagnostic assays in the 1960s and 1970s, a story used to circulate about the early days of assay development. The story was about an assay that was developed in the morning, transferred to manufacturing in the afternoon, and sent to the training lab that same evening. Suffice it to say that today assay development takes a little longer. The instrument systems and reagents are far more complex. In addition, government regulations require proof of many of the development results achieved at each stage of product development. Yet commercial managers still clamor for a shorter product development cycle.

This book focuses on data-analysis techniques that have been helpful in shortening the product-development cycle by solving problems in each phase of product development. Chapter 1 is an introduction that describes some serious quality problems in which patients have been injured and which have been caused by manufacturers, hospitals, or both. Reasons these problems escaped detection are explored. Chapter 2 covers the various environments of assay development—technical, commercial, and regulatory—and how many companies are organized to perform product development. The role of the consultant is also explored. Having the right solution is not always enough. One must be able to sell that solution to management, which at times may resist. Chapter 3 begins with the task of valuing a new opportunity. Financial modeling based on decision analysis is presented. Although decision analysis might not seem to be relevant to scientists and engineers, decisions based on these models greatly influence which products are funded. Scientists and engineers are actually well equipped to understand these models because they are similar in principle to data analysis based on scientific experiments. Chapter 4 describes adequate and not-so-adequate specifications. A poor specification can easily cause a product delay or worse—the wrong product. Many specifications could be improved by the remedies suggested here. Descriptions are provided for conjoint analysis, an important, quantitative marketing research method, and Quality Function Deployment (QFD), a method to improve setting specifications. Chapter 5 begins with some techniques for predicting when projects will be completed. It then describes some important aspects about the design of experiments that often are not covered elsewhere—namely, an experiment-planning checklist and advice about how to write a report that people will read. The chapter provides an overview of reliability growth management, a successful method that has shortened the time necessary to improve a system's reliability. Chapter 6 is about product evaluations. An error-modeling method is presented that is based on simulating assay results. The benefits of assessing total analytical error are explained. Multifactor protocols are described in detail. These are methods that allow information to be obtained in less time than one-at-a-time designs. The need to obtain information about outliers is explained, as are estimation and detection methods for outlier rates. Chapter 7 covers methods for improving the product after it has been commercialized. The ways in which the wording of performance claims can influence the rate of product complaints is examined. Data that are remotely gathered are analyzed by techniques such as the mean cumulative repair function and process capability measures.

Acknowledgement

Advances in science are facilitated by one's training, are based in part on previous work, and are often achieved with some level of collaboration. The author was fortunate to work in an environment that allowed one to be creative. He learned from Dr. Stan Bauer and Cuthbert Daniel at Technicon Instruments, from David Simmons at Corning Medical, from Larry Gentili at Chiron Diagnostics, and from many others. He benefited from a supportive environment provided by Dr. Richard D. Falb, at Ciba Corning Diagnostics. There were numerous collaborations with colleagues as well as many interesting problems to work on and techniques to learn at Technicon Instruments and the progression of companies in Medfield, MA.—Corning Medical, Ciba Corning Diagnostics, Chiron Diagnostics, and Bayer Diagnostics. The most rewarding collaboration was with Keith McLain, who developed the techniques described in the section on reliability while at Ciba Corning Diagnostics. Other fruitful collaborations have been with Brian Schlain, a statistical consultant, Pat Garrett, a clinical chemist now at Boston Biomedica, and Henk Goldschmidt, a clinical chemist from the Academic Medical Center of Amsterdam in The Netherlands. The author has benefited from the many collaborations with other industry members, hospital-based clinical chemists and pathologists, and government regulators in his long-standing relationship with NCCLS, a volunteer, consensus-based standards organization for laboratory medicine. Last, he would like to express gratitude to his wife, Ruby, for her work on this book and for her help and support throughout his career.

Introduction

Quality is in our hearts, minds, and spirits. But is it in the assay? Reflections on part of a diagnostic company's mission statement.

THE BUSINESS OF DIAGNOSTIC TESTING

In 2002, the business of in vitro medical diagnostics grossed more than $20 billion. That figure refers to manufacturers' revenues alone. Hospital and commercial laboratories generate additional revenues. As with any business, new technologies are developed that offer improved profits to the manufacturer, efficiencies and other desirable features to the laboratory user, and medical advantages to the clinician or patient. With new and existing technologies, both the manufacturer and the laboratory user have the responsibility to evaluate their products.

Quality Has Improved—Yet Physicians Rely More on Test Results

In the past 40 years, quality has improved greatly, largely because of automation. Over the same period, the practice of medicine has changed as well since assays play a larger role in clinical decision making. For example, an elevated cholesterol result might suggest the need for statin therapy, or an elevated PSA the need for a prostate biopsy. One must ensure that product evaluations keep apace with science to prevent problems from affecting the real end user: the patient. Sometimes they don't.

Lifescan: One Manufacturer's Problem

In 1993, Lifescan, part of Johnson & Johnson, began development of SureStep, its next-generation blood glucose meter for home use. Early in the development of SureStep, Lifescan knew that glucose readings over 500 mg/dL sometimes gave the message ER1 (Error 1) instead of the correct message HI. In addition, if a reagent strip was not fully inserted into the meter, the meter sometimes acted as if it was operating properly but would give an erroneously low reading.

In May 1994, Lifescan submitted an application to the FDA to market SureStep, failing to mention either that a high glucose reading could cause an ER1 message or that a partially inserted strip could give low results. The FDA approved the 510(k) submission in 1995. Lifescan began marketing the product in Japan, Canada, and the United States in 1996.

Almost immediately, Lifescan began receiving customer complaints. By 1998, the number of customer complaints had reached 2,700 and the U.S. government began investigating Lifescan with the help of two Lifescan employees. The complaints, which involved the ER1 problem and the strip-insertion problem, played a role in 61 cases of patient illness and injuries, including hospitalizations.

The U.S. government instigated legal action against Lifescan. In December 2000, Lifescan pled guilty to criminal charges and was ordered to pay $60 million in criminal and civil fines (*1*). In November 2001, Lifescan settled a class-action lawsuit related to this problem for $45 million (*2*).

University of Washington Medical Center: A Laboratory Problem

In February 1998, a 22-year-old woman consulted her doctor about irregular bleeding between menstrual periods. Tests revealed elevated levels of human chorionic gonadotropin (hCG), a laboratory assay used to detect pregnancy. This result was unexpected, because the woman was using contraceptives. Repeat testing failed to show the rapid upward climb of hCG levels that are associated with a normal pregnancy. Moreover, ultrasound and laparoscopy failed to detect any sign that the woman was pregnant. This led her doctors to suspect trophoblastic carcinoma, a rare and deadly form of cancer, which does respond to aggressive treatment when treatment is begun early. Chemotherapy was started, but because the hCG levels failed to fall, a hysterectomy was performed. The pathologists found no cancerous tissue. A subsequent full body scan showed suspicious spots on the patient's lungs, resulting in another surgery with removal of a portion of one lung. Once again, pathologists found no cancerous tissue. By this time 45 hCG tests had been performed, all of which had produced elevated results. Finally, a solution to this problem was found: samples were sent to a reference lab, which, using a different hCG assay, found that the patient's hCG level was normal. All of the previously performed tests were incorrect (false positives) because of an interference in the assay (*3*).

The woman sued both the test manufacturer (Abbott Laboratories) and the hospital and won the case for a total of 16.8 million dollars. The hospital suit could be viewed as a failure of the laboratory. The interference was caused by human antianimal antibodies, a known problem with immunoassays. Dr. Laurence Cole researched this problem in 78 cases. He found that 35 were false positives, and 12 led to unnecessary cancer therapy (*4, 5*).

Analysis of the Two Problems

Both problems appear much clearer with hindsight, which of course the participants did not have. These two problems are examples of real problems that occur during assay development and after assay commercialization. Neither problem would have been uncovered by a typical evaluation (i.e., those usually conducted to validate assays) that estimates imprecision and average bias, because both problems were related to outliers. Yet both problems caused considerable damage.

In the Lifescan problem, the frequency of both types of outliers, or "fliers" as they were called by Lifescan scientists, was very low. Moreover, to solve the ER1 problem, users could also check the color of the strip. The strip-insertion problem was caused in part by a user's failure to insert the strip fully, as stated in the product instructions. Because Lifescan knew about these problems ahead of time, it would be easy to ascribe this situation to corporate greed or a criminal act. In fact, Lifescan did plead guilty to a crime. Yet it is common during the development of many assays to worry that unforeseen problems may later cause an extremely small percentage of assay results to be outliers, which can cause considerable damage to patients. When such problems are uncovered, long discussions often follow about what to do, including the legal obligation to report problems (6). Perhaps more focus on outlier goals and estimation of outlier rates would have helped decision makers at Lifescan.

The problem at the University of Washington Medical Center can perhaps be characterized as a failure by clinicians to recognize that laboratory results can be incorrect. Additionally, knowledge transfer between the laboratory and the oncology departments failed to occur. Laboratory personnel would be much more likely than oncologists to know that human antianimal antibodies can interfere with immunoassays, yet most lab reports are provided simply as numbers without interpretation. See the Appendix for additional discussion.

WHY PROBLEMS OCCUR

It is naive to assume that all problems will disappear *if only. . . .* This book presents techniques that may help with the difficult problems such as the two discussed above. Conditions that inhibit the implementation of solutions for quality problems that arise during product development are also discussed. These topics are outlined here and are discussed in more detail in later sections.

Financial Incentives Don't Favor Allocating Resources to Quality

Chapter 3 explains that quality does not rank highly in the allocation of resources. This is understandable because the revenue returned by new or improved product features is almost always far greater than the revenue returned by improved quality. Additionally, time to market plays a key role in providing higher profits. This urgency puts pressure on managers to relax quality specifications to save time. The main incentive to allocate *any* resources to quality is either to prevent customer dissatisfaction and loss of revenue—which will happen if quality falls below some required level—or to comply with regulations.

Industry-supported Studies Favor Low-information-content Reports

Selling diagnostic assays is a highly competitive business. Advertisements and other product literature, such as package inserts, portray the product in the most favorable light. These materials are often prepared by marketing rather than technical departments. Often the performance claim material is abstracted from published evaluations. These studies, which are often conducted by hospital laboratories, are almost always funded by industry. This funding involves more than simply paying the hospital. An evaluation may involve the following division of work: The study protocol is provided by the manufacturer. The hospital provides samples and runs the assay. The data is shipped back to the manufacturer, whose scientists conduct the statistical analyses and write the report. Although financial and other assistance from the sponsor is acknowledged, the authors often do not include identities of the industry participants, an omission that creates the impression that the evaluation was conducted by an impartial third party. An example of low versus high information in a report is the use of a scatter plot rather than a difference plot. The scatter plot is primarily empty space. The difference plot better illustrates assay problems.

Corporate Culture

At times, data is available to help solve a problem, but the data or the methods necessary to analyze the data are not used. The behavior of senior management affects how an organization operates, even in this regard. Some years ago at a project review, a senior manager inquired about a project's status by asking, "How do you really feel about this project in your gut?" On a similar occasion, a consultant, brought in by senior management to speed up a stalled project, had been silent for the entire review. His sole question, which he asked at the end of the meeting, was, "Deep down in your heart, do you really believe this project is back on track?"

There is nothing inherently wrong with these questions. Yet if they are given too much emphasis, especially at the expense of *quantifying* results, then eventually the ability to answer questions on an emotional level will be valued more in the organization than the ability to provide answers through data analysis.

Inertia

Once a practice has been established, change is often slow. This is especially true in journal articles, in which authors present content based on previously published material (*7*). Journal instructions for authors and referees' reviews play a role in perpetuating practices. This may be one reason why scatter plots are still more common than difference plots (*8*).

HOW THIS BOOK CAN HELP

This book is written from the perspective of a manufacturer and from a data modeling and analysis view. It describes the development environment that exists in industry. Given the financial, technical, and regulatory forces that exist, sophisticated development and evaluation studies are often performed in industry but are rarely made public. The key studies are described here. Some of them can be performed only by manufacturers, others by hospital laboratories, and some by both. Besides covering the statistical aspects of performance evaluations, this book also covers data analysis and modeling methods that have been applied to financial modeling, development of specifications, instrument system reliability, and other applications.

The book is also written from the perspective of a consultant. This is relevant not just for consultants, but for anyone who has an idea and tries to implement it. Just because one is right does not mean that one's solution will be used. A famous example is of a 200-year delay for the British navy to adopt measures to prevent scurvy, despite evidence that a solution had been found (*9*). One must be able to identify, understand, and deal with the resistance that often occurs when new solutions are suggested. Thus, "That's a great idea. Let's do it on the next project" is a common response to a consultant's proposal. Strategies for successfully implementing solutions are explained, as are signals that indicate when it is time to stop trying and move on.

REFERENCES

1. http://www.usdoj.gov/usao/can/press/html/2000_12_15_lifescan.html. Accessed 2 March 2002. This summary has a link to the plea agreement from the United States District Court, Northern District of California.
2. *San Jose Mercury News.* "Mintz H. Lifescan Settles Allegations over Diabetes Home-Testing Devices." 27 November 2001.
3. Sainato D. How labs can minimize the risks of false positive results. Clin Lab News 2001;27(1):6–8.
4. Rotmensch S and Cole LA. False diagnosis and needless therapy of presumed malignant disease in women with false-positive human chorionic gonadotropin concentrations. Lancet 2000;355:712–715.
5. Cole LA, Rinne KM, Shahabi S, and Omrani A. False positive hCG assay results leading to unnecessary surgery and chemotherapy and needless occurrences of diabetes and coma. Clin Chem 1999;45:313–314.
6. *http://www.fda.gov/cdrh/fr/fr1211t.html* accessed 2 March 2002.

7. Hackney JR and Cembroski GS. Need for improved instrument and kit evaluations. Am J Clin Path 1986;86:391–393.
8. Dewitte K, Fierens C, Stöckl D, and Thienpont LM. Applications of the Bland-Altman plot for interpretation of method comparison studies. A critical investigation of its practice. Clin Chem 2002;48:799–801.
9. Mosteller F. Innovation and evaluation. Science 1981;211:881–886.

Appendix
The Role of Assay Error in Decisions to Approve Assays for Use

In the case of the University of Washington Medical Center, the clinicians and the laboratory were using the hCG assay to aid in the diagnosis of cancer. However, this hCG assay was not approved by the FDA for this use. This raises the question: "What choices must be made when deciding whether to allow an assay to be used?" These choices include the following:

1. Allow the assay to be used, knowing that some injury and death may occur from possible assay errors.
2. Don't allow the assay to be used, knowing that some injury and death may occur as a result of the absence of information the assay would have provided.

Injury or death attributable to assay error will almost always cause more public concern than injury or death attributable to lack of information that would have been provided by an assay, assuming that each cause could somehow be proven.

These issues will always weigh on regulators' minds. Yet if an assay has the potential to reduce injury and death because of the information it provides, then it is the job of manufacturers and hospital labs to reduce assay errors—especially the large errors that contribute to incorrect treatment decisions.

2

The Diagnostic Assay Development Landscape and the Role of Consultants

That's a great idea. Let's do it on the next *project.* An indication that a presentation about implementing an idea is not going well.

The modern diagnostic assay environment began in 1957 with the introduction of the autoanalyzer (*1*). A significant consequence of this invention was automation of the difficult manual steps required to run an assay. Automation improved quality and spurred demand because the cost of assaying samples was reduced.

THE TECHNICAL ENVIRONMENT

Perhaps the biggest challenge in assay development is not mastering any single technology, but rather integrating several complex and disparate technologies. A typical assay that runs on a current analyzer system contains

- a hardware platform with electromechanical, optical, and fluidic systems
- software, including hardware control systems, user interfaces, complex algorithms that convert raw signals to concentrations and test for result quality, data management, and connectivity protocols
- biochemical reagents that are sensitive to various environmental factors

To remain competitive, management must develop and often revise technology strategies based on studying technology trends. Some newer technologies turn out to be important, and others less so. For example, automation achieved through hardware and software has continued to be an important technology, whereas expert systems have not yet played the role once forecast for them. Research and development teams are always under pressure to shorten the product development cycle while providing new product features and maintaining or improving product quality.

THE COMMERCIAL ENVIRONMENT

The in-vitro medical diagnostic industry can be grouped by discipline into areas such as clinical chemistry (also called routine chemistry), hematology, microbiology, immunoassays, nucleic acid and gene probes, blood gas, and many, smaller segments. Sample types include serum, plasma, whole blood, urine, and other body fluids. Many of the products are commodity-like, low-profit margin items such as a routine assay for uric acid or cholesterol on a large analyzer. Higher-profit margin products include new tumor markers or other tests discovered through medical research, and probe assays that are more sensitive with lower detection limits. Although there are many patents, they rarely confer the same profit protection from competitors that is seen in the pharmaceutical industry. The market segments include large national reference labs, hospital labs, point-of-care testing within the hospital, doctors' offices, and home testing.

Assays in all market segments are designed to meet the needs of primary and secondary customers, who include

- people who perform the assay
  - technicians in hospitals, reference labs, and doctors' offices
  - nurses, doctors, and technicians in point-of-care settings
  - home users
- people involved in the buying decision
- clinicians and patients
- regulatory agencies

THE REGULATORY AND MEDICAL ENVIRONMENT

Assays are regulated in most countries. Different levels of regulation are required for different assays. In the United States, a routine assay such as that for cholesterol requires data that show that the new assay is "substantially equivalent" to an existing, approved assay. New markers and some assays based on new technologies require much more extensive studies for approval. All of the regulations are affected by medical trends, with physicians participating in the formulation of most regulations. Implicit in these regulations are the notions that assays must lead to improved patient care and that assay quality must be good enough so that assay error will be small enough to be unlikely to contribute to incorrect medical decisions.

THE MANAGEMENT ENVIRONMENT

Assays are developed within a centralized or a decentralized management structure. Because a company by definition consists of only one entity, it always involves some form of centralization, such as the President or the CEO and the top level of management. Some staff groups such as human resources, finance, regulatory, and others are also often centralized. Major departments that might be decentralized are commercial divisions—such as marketing and sales—and R&D. If R&D is decentralized, it usually reports to a decentralized commercial division.

A centralized group does not necessarily perform all its activities in one location, even when different business segments are located in different geographic areas. Some centralized groups operate by maintaining a staff at each local site, with the local group supporting the needs of that site. The function is nevertheless considered centralized because all local groups are managed by one person.

The Five Stages of Product Development

Whereas companies are organized differently, many companies follow, formally or informally, a stage gate process that breaks down development of new products into a series of steps (2, 3). The "gate" in stage gate means that a go/no-go decision is needed to advance the project to the next stage. Often the next stage is started with an automatically granted go decision (except for the product launch stage), yet the breakdown into stages is useful for describing the different activities that occur.

The five stages are as follows:

Stage I: Researching new opportunities—In Stage I, companies decide which opportunities to fund by valuing projects financially and by assessing the probability of technical success.

Stage II: Proving feasibility—The project has been funded. Specifications (often started in Stage I), are created to provide scientists and engineers with the requirements they need to design the product. At the end of Stage II, evaluations are performed to determine whether feasibility has been met.

Stage III: Scheduled development—Most product development occurs in this stage, with the two major efforts being assay development and instrument development. System integration is done to ensure that all parts of the system, often developed separately, work together on the instrument. Regulatory approvals are submitted, a Stage II activity for some products.

Stage IV: Validation—Evaluations at customer sites—often started in earlier stages—are performed with final or nearly final assays. Extensive internal evaluations and process validations are performed, including software validation.

Stage V: Commercialization—The product is positioned in the marketplace through advertising campaigns, customers are trained, and product warranty claims are issued. After product release, product development continues by implementing product features that have been deferred, adding new product features, and resolving product complaints.

These steps will be discussed in detail in subsequent chapters.

THE CONSULTANT'S ENVIRONMENT—HOW CONSULTANTS GET THEIR SOLUTIONS IMPLEMENTED

Much of the remainder of this book deals with data analysis and modeling techniques used throughout the product development cycle. Often, optimal techniques are not yet in place. The following section discusses how to get one's expertise used. The information in this section applies conceptually to anyone who is trying to sell and implement ideas (4).

How Statisticians Are Often Perceived and What They Really Do

Statisticians in the medical diagnostics industry often have to deal with less than desirable perceptions about their function. Oftentimes, "crunching the numbers" is perceived to be the statistician's main role. Yet this is only part of what most statisticians do—or at least should do. Common requests to a statistician include "What should my sample size be?" or "What do my data mean?" A typical response from a statistician to these requests is "What is your goal?" An example follows.

A mixup occurred in the production of a quality-control material for a TSH assay. For an unknown period of time, labels for concentration level 1 were mistakenly put on vials for concentration level 3. Customers using the mislabeled vials would believe that their assay was out of control, when in fact the control material was incorrectly labeled. As soon as the error was discovered, the "bad" vials were discarded. However, no one could be sure that all the vials that should have been discarded, actually were. The lot size for each concentration level was 1,500 vials. The vice president in charge of quality asked a statistician to analyze data from an experiment in which 20 concentration level 3 vials were to be opened and assayed. The vice president suggested, "If no bad vials are found, we can be pretty sure that no more mislabeled vials are present."

This example typifies a common problem. The statistician asked the vice president in charge of quality what he meant by "pretty sure." After some discussion, it was agreed that the vice president wanted to be 99 percent confident that the population of vials contained 99 percent good vials. With calculations based on the hypergeometric distribution, the number of vials required to be tested (destructively) was so high that it was determined to be more economical to destroy the entire lot.

The point of this example is that the output provided by the statistician went beyond the original request. This is often the case. One can also review the sequence of events. A problem occurred, so a meeting was called to discuss it—in this case, without including the statistician. A decision was reached about what to do, and the statistician was contacted to perform a task. The statistician provided insight into the problem, which led to a different task. One can then ask why the statistician is sometimes not included in the original meeting.

Some managers are relationship oriented rather than technically oriented. For these managers, a problem is viewed more with respect to relationships rather than to technical issues. Key managers, chosen to attend the meeting based on relationships, propose solutions to the problem and assign implementation of their solution. If statisticians develop sufficient relationship skills, they will be included in initial meetings.

When managers are technically oriented, statisticians may face different challenges. Technical managers have often developed expertise in their respective fields without much formal use of statistics, at least in technical fields related to developing diagnostic assays. Again, key managers meet and, in this case, experiments are proposed without the statistician, who is relegated to "crunching the numbers." If the statistician can develop a track record of accomplishments, he or she will likely be included in the meetings that are used to design experiments rather than simply be asked to "crunch the numbers."

The previous section covered issues a statistical consultant may face when asked to help with a problem. The next section provides implementation suggestions when someone has an idea to solve a problem but is not being asked to help. The "someone" could be a consultant, or any scientist or engineer who is trying to implement an idea.

When Management Resists—Techniques Used by Consultants to Implement Solutions

A person has an idea to solve a problem. This person knows that:

- the problem is an important one for the organization
- his or her solution will probably solve the problem
- management is willing to listen to the proposal, but implementation will depend on the person's ability to sell the solution

In some cases, management approves the proposed solution without much discussion. However, this outcome is rare. Understanding successful selling strategies requires insight into the reasons management may resist proposed solutions. It seems illogical that management would resist a solution that would most likely solve an important problem. Yet, there are reasons. Solutions must always be approved by management, even when the decision maker has no apparent management responsibility. An individual contributor who has few or no subordinates still has an implied management role since he or she may control millions of dollars of company funds or play a crucial role in the product development process.

It's About Control

Many managers, including the individual contributors mentioned above, have a certain amount of power or control in their organization that they will go to great lengths to protect or acquire. Solutions suggested by consultants almost always threaten the control held by these managers.

Sometimes a manager may not completely understand the technical details of a solution, so the way the actual solution unfolds—including some decisions—may either be made by the consultant or be shared between the manager and the consultant. Managers who can comfortably acknowledge their lack of expertise or those who view the consultant as just another subordinate have no problem with this situation, albeit for different reasons. The managers who cannot, fear that others will perceive the situation as a loss of control or power. The reality is that in some cases, the loss of control is needed to implement the proper solution and ultimately improve profitability.

Discussions about relinquishing control almost never occur. Instead, managers may offer one or more forms of resistance to prevent the project from taking place. This resistance includes comments like the following:

- "Great idea, let's do it on the *next* project."
- "I understand what you're saying, but it simply won't apply here; our situation is different."
- "I just don't understand what you're saying."
- "It's too [complicated], [expensive], [will take too long], etc."
- "I once took a course in statistics."
- "Yes, let's do it."

Regrettably, the last item in this list can be a form of resistance, too, and causes one to have to distinguish whether *yes* means *yes* or *no*. One technique to distinguish what *yes* means in this instance is to plan the first implementation step at the original meeting. If there is reluctance to plan this first step, then agreement may actually be a form of resistance.

The strategy to deal with resistance depends on the personalities involved. Some examples are presented below. Be aware that however plausible an implementation strategy is, it may be unsuccessful. The consultant must at some time know when to move on to another project.

Identify and directly address the stated resistance—Example: "Your main concern seems to be fear that time to market will be longer. We can show you three examples where these techniques actually shortened the time to market for similar projects."

Deal with the control issue—Example: "Once we have trained you and your staff, you will be able to perform these techniques by yourself. We will transfer all the technologies and move on to something else. You will be recognized for having provided these improvements."

The technical rationale—Scientists and engineers have been trained to solve technical problems and most likely are quite successful at it. Nonetheless, it is exceedingly difficult to try to help people who are convinced that they don't need help. Also, many scientists are trained to work independently and not in teams. Finally, there is an additional factor. An assay or an instrument's lead development scientist or engineer, and not the consultant, will be held accountable for any design failures that might occur, and this must be respected as a legitimate concern. A scientist or an engineer who is unfamiliar with a new technology will be reluctant to use it and will rationalize that, in the event of a design failure, he or she does not want to be in a position to try to justify a design choice that was based on a technique he or she does not understand. Training can help to alleviate these concerns. Besides lowering resistance to the use of new methods, training has many other advantages. It can be a core part of a learning organization. One goal of training suggested by reliability consultant Dr. Ralph A. Evans is "to make it easy to do the right thing and hard to do the wrong thing."

The Successful Consulting Cycle

In a typical successful consulting cycle, the consultant has overcome resistance to getting a technology used. This may occur during the first attempt at implementation or during subsequent attempts. Suddenly the recipient not only "sees the light" but masters the technology and begins using it without any further need for the consultant. He or she gives management presentations that promote the benefits of the technology and during these presentations, often gives little or no credit to the consultant. The recipient has taken ownership of the technology and has internalized it within his or her department. Lack of credit given to the consultant is expected as part of the successful consulting cycle, and the internal consultant should have taken steps to ensure that management is aware of the consultant's true role.

Technology Transfer—The Benefits of a Learning Organization

Modern technology is extremely complex, with advances occurring frequently. These frequent changes make it difficult to keep up with developments in one's own field, let alone in other fields. However, many new products or techniques often result from technology being transferred from one area to another (5). Using "best practices" often shortens development time and improves product quality. Some statistical examples of knowledge transfer include the following:

- ROC curves used to assess diagnostic accuracy, adapted from signal detection work in radar
- Reliability growth management used to improve reliability and shorten development time, adapted from the defense industry

- Levey-Jennings charts used to monitor assay quality, adapted from industrial quality programs

The benefits of technology transfer may be hard to quantify, but as discussed in Chapter 1, the lack of knowledge transfer between the pathology and oncology departments at the University of Washington Medical Center may have contributed to the problem they experienced.

Technology transfer within companies is often enabled by interaction among diverse divisions. Often the knowledge itself is not new, but it is new to the group acquiring it. There is also a spectrum ranging from acquiring and using the knowledge in its current form to considerable modification of the original ideas.

Knowledge transfer, while conceptually attractive, does not occur automatically. In fact, it may not occur at all in the absence of a program that recognizes the organizational and technological requirements needed for encouraging organizational learning. Examples of techniques or programs include the following:

- A list of people who have knowledge in key areas is valuable and different from an organizational chart, which reflects power instead of knowledge.
- Management incentives could make the difference between scientists and engineers sharing knowledge rather than hoarding it.
- Ideas that are valued for usefulness rather than originality will encourage adapting existing work rather than relying solely on new ideas.
- Corporate intranets can be developed to use knowledge-based software that facilitates learning.
- Colocating organizational groups or centralizing functions may facilitate knowledge transfer.

Training as Part of a Learning Organization

Acceptance of new technology often passes through three phases:

1. Demonstration of theory
2. Field testing
3. Acceptance by users

In diagnostics, evaluation methods based on data analysis follow the above three phases. A method's theory is published in statistical journals. Field testing results are published in applied journals such as *Clinical Chemistry*. User acceptance often depends on the quality of the training.

Training can be conducted by having regularly scheduled courses. However, a particularly effective way to conduct training is to make it project specific. That is, training is tailored to a specific project and conducted at the appropriate time in the project's development cycle.

REFERENCES

1. Lewis LA. Leonard Tucker Skeggs — A multifaceted diamond. Clin Chem 1981;27:1465–1468.
2. Cooper RG. Winning at new products. Reading, MA: Addison-Wesley, 1986.
3. Cooper RG. Third-generation new product processes. J Prod Innov Mgmt 1994;11:3–14.
4. Block P. Flawless consulting: A guide to getting your expertise used, 2nd ed. New York: Jossey-Bass, 1999.
5. Davenport H and Prusak L. Working knowledge: How organizations manage what they know. Cambridge, MA: Harvard Business School Press, 1998. See *http://www.cio.com/archive/021598/excerpt.html* for an excerpt. Accessed 2 March, 2002.

3

Stage I: Researching New Opportunities

There's a window of opportunity provided we can get the product out on time. —
A common statement heard at management meetings early in product development.

The goal of most commercial entities is simple: to be profitable. The goal of profitable entities is also simple: to be more profitable. Most issues within a company can be related to their impact on profitability, as the following questions illustrate:

- Which combination of projects should be funded, and at what level?
- Should a new project get funding at all?
- Should a new product feature be added that delays product launch?
- Will a new technology be successful?
- Is the new product good enough to launch?

This chapter considers how these and other issues can be addressed within a decision-analysis framework, using financial models. Nonprofit organizations (such as nonprofit hospitals) must also understand financial models. Nonprofit organizations that do not recover their expenses are forced to cut services.

WHY SCIENTISTS AND ENGINEERS NEED TO UNDERSTAND FINANCIAL MODELS

New projects require resources, and there are seldom enough resources to support all new proposals and existing projects. Thus, the opportunity to work on a project may depend on a scientist's or an engineer's ability to sell management on the merits of the project. A presentation requesting resources that promotes a product's new features or its improved quality will not be received as well as one that shows an attractive Net Present Value (NPV) analysis of cash flow. The presentation about improved quality will be politely listened to, whereas a presentation that forecasts improved profitability will generate excitement. Money is the language, and financial modeling the grammar, of senior management, so scientists and engineers must learn this language. They are actually well equipped to do so, since the techniques used in financial modeling and analysis are often similar in principle to those used in assay development and evaluations. One gets input about parameters that have uncertainty, analyzes the data according to a model, and predicts results about future events. A major difference in financial analysis is that the uncertainty in the input data is about future events, not past events.

USING DECISION-ANALYSIS–BASED FINANCIAL MODELS TO VALUE OPPORTUNITIES
Decision-analysis Background and Terms

Modern decision analysis used in financial modeling is based on work by Raifa (*1*) and by Matheson and Howard (*2, 3*). The latter two formed Strategic Decisions Group (SDG), one of the best-known consulting firms in this field.

Terms used in decision analysis and financial modeling include the following.

10-50-90—A means of soliciting the probability of uncertain events. In this scheme, "10" represents a 1-in-10 chance that the parameter value could be as low as some amount, "90" represents a 1-in-10 chance that the item could be as high as some other amount, and "50" is the most likely value for the parameter. See the Appendix for how this is used in NPV calculations.

Cash flow—Cash flow is the cash the company has on hand to pay bills. Profitable companies with a negative cash flow can go out of business, which is why cash flow rather than profitability is the metric used in financial modeling.

Condition—A variable or factor that affects the output. For example, the launch date for a product almost always affects, or "conditions," the NPV.

Discount rate—The interest rate assumed for future borrowing.

Expected NPV—The probability weighted average of a distribution of NPVs. See Appendix for a calculation example.

Frame—The context of an opportunity. An example of a frame is, any in-vitro diagnostic assay that uses whole blood as a sample.

NPV—Net Present Value. The value of future cash flows expressed in today's dollars.

Decision analysis requires a certain infrastructure to be in place before models can be performed for a series of opportunities. The required tasks include the following:

- selecting decision-analysis software
- creating a decision-analysis team
- preparing an influence diagram
- defining inputs that are required
- building financial equations based on the required inputs.

Once these steps are completed, the actual decision-analysis model, which involves the following, can be performed for a series of similar projects:

- soliciting the input data
- performing the analysis
- reporting results

Selecting Decision-analysis Software

Decision-analysis software should be easy to understand by all persons involved in the financial modeling. Ideally, the software would be a spreadsheet add-in or at least compatible with a spreadsheet. The Decision Analysis Society, a subdivision of INFORMS (Institute for Operations Research and the Management Sciences) maintains a Web site for decision analysis software, reading lists and other aids at *http://faculty.fuqua.duke.edu/daweb/dafield.htm.*

Creating the Decision-analysis Team

In traditional financial modeling, there is often no multidisciplinary team devoted to an NPV analysis because the finance department performs all modeling and solicits individual input from the person with the information required for the model. In the team approach, input is requested from this person so that anyone on the team can comment on it. This allows for checks and balances to garner higher-quality input.

Preparing an Influence Diagram

A simplified influence diagram for a typical assay is shown in Figure 3-1. In this diagram, the lines pointing to the top-level event—NPV of cash flows—originate from the factors that influence it. The goal of the influence diagram is to show the relationships between areas that need data input as well as the level of detail required. Thus, an influence diagram sets the level of granularity required for the data. For example, administration costs for most new projects will simply be a fixed percentage and don't require any more detail. On the other hand, manufacturing costs might well require considerable detail for a project that involves new manufacturing techniques, so this section of the influence diagram would have be to be expanded. Choosing the level of detail is important because too little detail will result in a poor model while too much detail will slow the process by being overly complicated and costly.

Given an influence diagram, the next step is to prepare a decision tree. A decision tree lists all of the inputs required. The actual preparation of the influence

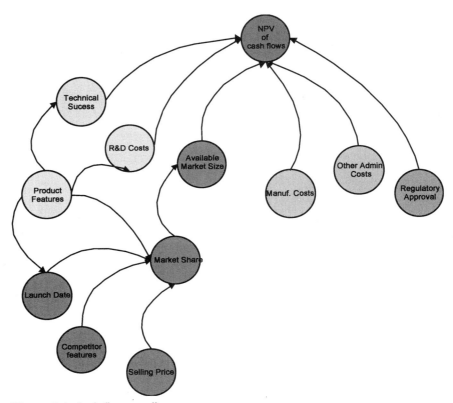

Figure 3-1 An influence diagram.

diagram or especially of the decision tree is usually performed with software. As an example of data required for input, consider the launch date, which is almost always a controversial parameter. For many people in product development, there is a single launch date—the official launch date, which is treated as a constant regardless of how many times it is changed. Yet, typically in decision analysis, the launch date is treated as an uncertain event and often values for three launch dates are requested according to the 10-50-90 rule. A fragment of a decision tree is shown in Figure 3-2, for the launch date example. The optimistic "10" case corresponds to a launch date that may be as soon as 20 months (e.g., this event corresponds to a 1-in-10 chance). Associated with this input is the effect of early launch, in this case a 30 percent market share. For the pessimistic "90" case, there is a 1-in-10 chance that the product may not launch until 60 months, which will result in a market share of 15 percent. The most likely launch date is 30 months, resulting in a market share of 25 percent.

Inputs for other variables would also be provided so that a complete list of required variables is prepared. Note that in this process, the math for a financial

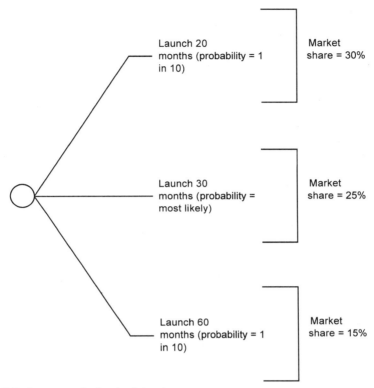

Figure 3-2 A segment of a decision tree.

model is being developed, part of which is implied in Figure 3-2 (albeit in an oversimplified version since other factors would also influence market share). That is, the basic equation—profit equals revenue minus cost—is expanded based on the decision tree. The actual complete financial model, while simple in principle, can have hundreds of variables. The tree format is also an ideal way of expressing the many "if-then" branches.

Techniques to Solicit Unbiased Data

As with any modeling activity, the quality of the output depends in part on the quality of the input. The team approach is used to help ensure high-quality inputs through consensus. A key success factor in this process is the facilitator who manages the input session. Detailed discussion of facilitator techniques is beyond the scope of this section; many companies have facilitators on staff, who need not be experts in the technical area. Their function is instead to ensure that inputs are balanced and appropriate. For example, a dominant personality with no real knowledge about a parameter might nevertheless try to convince the group about its value. It is the facilitator's role to encourage participation from the person who does have expertise.

As another example, a marketing manager might suggest that a proposed new opportunity will garner 60 percent market share in the first year. However, if similar products from the company have previously achieved 10 percent market share in the first year, then the proposed claim for 60 percent requires more justification. Ideally, all the inputs are supported by some type of justification, often using historical data. It is also important that this justification be recorded in writing. These models are not isolated, one-time events; they are continually updated throughout the life of the project. New information can change the values for the parameters. A good practice is to keep all model versions (results, input data, justifications) rather than overwriting older versions with newer ones.

Performing the Analysis—The Results

The Base Case

Typically, several analyses are performed to complete a decision-analysis model. A base-case result is achieved by running the analysis while holding all parameters at their most likely values. Reports provided by this analysis include the total NPV across a specified period such as 10 years and year-by-year charts and tables for variables such as revenue, market share, and cost.

Sensitivity Analysis

Sensitivity analysis is performed to assess risk. A one-way sensitivity analysis works by calculating NPV for each value of one parameter while holding fixed all the other parameters at their most likely (base) values. This is continued for all parameters. The results are sorted by the parameters that show the biggest NPV range between the parameter's high and low values. Figure 3-3 shows an example of the sorted results, displayed in a chart called a Tornado diagram. In this example, the top bar, "launch date," has the greatest risk to profitability. If the late launch date rather than the early launch date is realized, profitability will be reduced by $8.5 million. Each of the other parameters is listed in its decreasing order of risk to profitability. At the bottom of the figure (not all parameters are shown), "sales cost" has a much lower risk on profitability. Sensitivity analysis helps management focus on high-risk issues. In the current example, decision makers could allocate more resources toward ensuring that the launch date is met rather than reducing the sales cost.

Note that risk to profitability is not the same as influence on profitability. Thus, a launch date may have the highest risk to profitability, and average selling price may have a negligible risk to profitability, only because the

NPV from Product Life Cash Flow
Base Value: $11,776.28

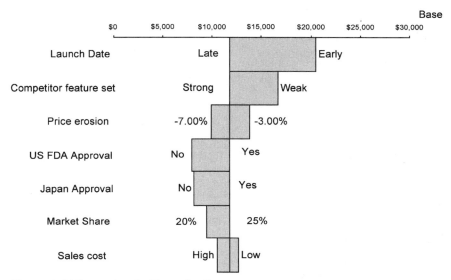

Figure 3-3 Example of a Tornado diagram.

average selling price range may be very small. However, the selling price will always be an influential factor in profitability.

Be aware that a non-decision-analysis–based financial model can also contain a sensitivity analysis. The problem is that this analysis is usually performed without any use of probabilities. One simply takes the base case variables and adds and subtracts a percentage (often 10% or 20%). Sensitivity analysis then proceeds as above for the decision-analysis case. The problem with not using probabilities can be illustrated by using the launch date example again. If a launch date of 30 months is used and is overly optimistic, it might be equivalent to a 25 percent likelihood (remember that in a non-decision-based model, there is a single value for each parameter and no probabilities). Taking ± 20 percent of this 30-month launch date give dates of 24 and 36 months, which in this made-up example might be associated with probabilities of 5 percent and 50 percent. This coverage of 5–50 percent will underestimate the true risk of the launch date variable.

Distribution Analysis

In a distribution analysis, the top three to five parameters from a sensitivity analysis are used to compute NPVs for all possible combinations of these

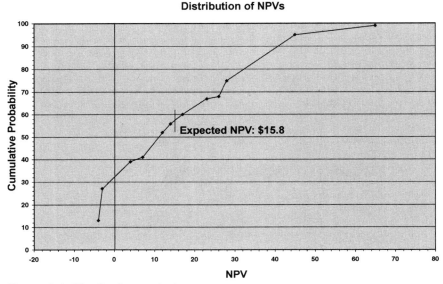

Figure 3-4 Distribution analysis.

variables. For example, if four parameters are selected, each of which has three levels, then 81 NPVs will be calculated. An example of a distribution analysis chart is shown in Figure 3-4. The distribution of NPVs is different from a distribution of a diagnostics assay parameter such as bias in patient samples. For example, one could calculate bias from reference for each of many samples and display the results in a graph similar to that in Figure 3-4. The graph would be a display of events that *actually* occurred even though the specific location of a particular point on a graph such as Figure 3-4 could not be predicted. The graph would imply that future results would be contained in a similar graph. For NPVs, the situation is different. The distribution of NPVs is based on assumptions about *future* events. When the actual NPV event occurs, it will be a single point and possibly not even within the range of NPVs in the distributional analysis.

TECHNIQUES TO IMPROVE DECISION-ANALYSIS MODELS

The Use of Options

A traditional financial equity call option allows the purchaser to obtain a stated number of shares of a company (the number of contracts) at a specific price (the strike price) by a certain date (the expiration date) by paying a stated amount (the premium). Although still a controversial topic, the use of real options has

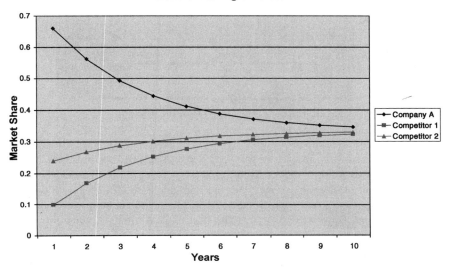

Figure 3-5 A simple Markov analysis.

been used to account for additional value beyond that estimated by traditional NPV calculations. The additional value is implied in management's ability to change decisions (e.g., their option) about new-opportunity funding. A Web site provides an introduction to this topic (*4*).

Markov Analysis

Most diagnostic assays are sold based on the ability of the sales department to retain customers or win new customers from the competition, with market share either gained or lost to competitors. Markov analysis is a technique that helps predict market share for this case. Figure 3-5 shows the output of a simple Markov analysis. In this figure, three companies each have 80 percent customer retention, whereby 10 percent is lost to each of the other two companies. Company A starts with 80 percent market share, competitor 1 at 0 percent market share, and competitor 2 at 20 percent market share. The figure shows that, with these input parameters, all companies approach equal market share as a function of time.

Markov analysis works by assuming a series of states and transitions. A simple example follows.

In this example, there are two diagnostic companies, Apex Diagnostics and Zenith Diagnostics. Apex Diagnostics retains 90 percent of its customers each year and loses 10 percent to Zenith Diagnostics. Likewise, Zenith Diagnostics retains 85 percent of its customers each year and loses 15 percent to Apex Diagnostics. Figure 3-6 is referred to as a state transition diagram and is used to

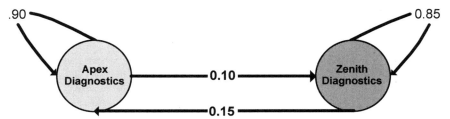

Figure 3-6 An example Markov state transition diagram.

portray the above situation. Thus, for Apex Diagnostics, each returning customer either purchases a new product from Apex (shown by the arc labeled 0.90) or switches loyalty to Zenith (shown with the line labeled 0.10). Based on the diagram, a transition matrix is constructed (Table 3-1), which contains the transition probabilities implied by the diagram. The sum of each row is 1, since probabilities must add to 1.

Currently, Apex Diagnostics has a 35 percent market share, with the remainder held by Zenith. If these brand loyalties remain constant, Markov analysis can be used to predict future market shares. The current market share states are reflected in the state matrix (Table 3-2).

Markov theory says that in year t, the state of the system (S_t) is calculated as follows:

$$(S_t) = (S_{t-1})(P) = (S_{t-2})(P)(P) = \cdots = (S_1)(P)^{t-1} \tag{3-1}$$

where S_t = the matrix of the states of the system at time period t
P = the matrix of probabilities for transitioning between states
t = the time period

One can set up these calculations in a spreadsheet, since $(S_2) = (S_1)(P)$ and $(S_3) = (S_2)(P)$, and so on. Note that multiplication order is important when using matrix multiplication.

Table 3-1
Transition Matrix P

	To state Apex	To state Zenith
From state Apex	0.90	0.10
From state Zenith	0.15	0.85

Table 3-2
State Matrix S

Apex market share (%)	Zenith market share (%)
35	65

Table 3-3
Market Share Results from Markov Analysis

Year	Apex (%)	Zenith (%)
1	35.0	65.0
2	41.3	58.8
3	45.9	54.1
4	49.5	50.5
5	52.1	47.9
6	54.1	45.9
7	55.6	44.4
8	56.7	43.3
9	57.5	42.5
10	58.1	41.9

Table 3-3 shows the results of years 1–10 for this example, provided that loyalties remain the same during this period.

This example illustrates that a small percentage difference in brand loyalty can override a lower starting market share. Although this may be intuitive, it is helpful to quantify results. Although constant brand loyalties were used, they could be set to change from year to year. If that were the case, variable P would also have the same time subscript as variable S in Equation (3-1).

The output of a Markov analysis is often the input for the market share variable in a financial analysis.

Methods to Evaluate the Probability of the Technical Success of Opportunities

New ideas come from a variety of sources, such as

- University research
- Basic or applied research within the company
- Basic or applied research from another company

In any of these cases, money will have to be spent to develop the new technology through licensing, support of development inside or outside the company, or a combination of these funding types.

Two main questions need to be asked about the probability of technical success for a new opportunity: Will it work? and if so, When will it be ready?

Different States of Knowledge Require Different Strategies

The best method that can be used to answer the above questions depends on the available state of knowledge for the technology.

High State of Knowledge. In this situation, knowledge can be expressed with mathematical equations based on physical properties. For example, one would not need empirical experiments to determine how much substrate to add to an enzymatic reaction if it followed Michaelis-Menton kinetics. One could simply calculate the amount. In many cases, knowledge will not be perfect, but it will still allow equations to be used to help assess probability of technological success. Thus, a noninvasive glucose assay based on near-infrared is governed by the laws of spectroscopy.

Medium State of Knowledge. In the medium state of knowledge, knowledge can be expressed by empirical equations based on experiments. An analogy might be mapping a mountain range with simple survey tools rather than with the use of satellites. One has to perform experiments to acquire sufficient knowledge. As an assay example, one might perform an experiment to determine whether a dose-response curve is steep enough to justify further assay development.

Low State of Knowledge. In a low state of knowledge, each experiment provides information but does not seem to form a pattern that allows empirical equations to be developed. For example, the theory to find an optimal surfactant for an assay might be insufficient, so each of many surfactants is tried.

One must choose a technology evaluation method that matches the state of available knowledge. One must also be aware that people will solve problems with the skills they possess, and it is rare for people to be skilled in both theoretical methods (high state of knowledge) and empirical methods (medium state of knowledge). The skills for an evaluation technique may not be available on site. Therefore, it is important for management to ensure that the wrong technique is not used because of a lack of in-house knowledge.

RESULTS BASED ON DECISION ANALYSIS

Why Management Always Wants the Product Released Sooner

Most new in-vitro diagnostic assay products contain new features and command an initial higher selling price when released and perhaps an increase in market share. This results in higher profit. However, because the diagnostic assay industry is highly competitive, the higher selling price, and perhaps the higher market share, will not be sustainable. As near-commodity status of the product returns, profit will be lower. Although management does not need decision analysis to make the above arguments in order to justify driving the organization to release products sooner, use of decision analysis quantifies the benefits of an earlier release as well as the penalties of a later release.

Returning to the launch date example in Figure 3-2, we see three possible launch dates for a new product, an optimistic, persimistic, and the most likely date. In some companies, the only date that most project members will see is the optimistic date (corresponding to 20 months). It is often a management strategy to challenge the organization by providing "stretch" goals. Moreover, these motivational goals are often provided as being "cast in concrete" as a fixed date. All will be well (we shall not debate the merits or perils of stretch goals), as long as one does not abandon the three dates and their associated probabilities for financial modeling. This is because modeling both identifies launch date as an important financial risk to NPV of cash flows and quantifies the impact of this risk. This allows for the development and possible deployment of contingency plans. Thus, as events unfold and a project is delayed, contingency plans might provide more funding to try to shorten the launch date, or perhaps to actively promote an older product longer or a project that gives the older product a facelift and relaunches it.

Why Quality Ranks Low in Terms of Financial Rewards

New opportunities are valued for their potential to be profitable. A Kano diagram (Figure 3-7) helps illustrate why quality is never highly regarded in terms of profitability.

In this diagram, the straight line expresses the conventional wisdom that providing more features is rewarded with increased sales that are proportional to the amount of new features. An example of this might be a proportional rise in customer value as the number of assays increases. The top curve shows that an unexpected yet highly desirable feature, such as the original automation in diagnostic instrumentation, provides explosive revenue growth. The bottom curve is associated with some product features, such as accurate answers and other forms of quality, that are expected by the customer. If more quality is provided, there is only a small gain in revenue, if any; however, if the level of quality drops below a certain amount, there is a large and often disastrous drop in revenue associated with customer dissatisfaction.

Portfolio Analysis

In most companies, more projects are proposed than there are available resources to support them. The goal of a portfolio analysis is to provide a means to allocate resources among competing projects. If one has conducted

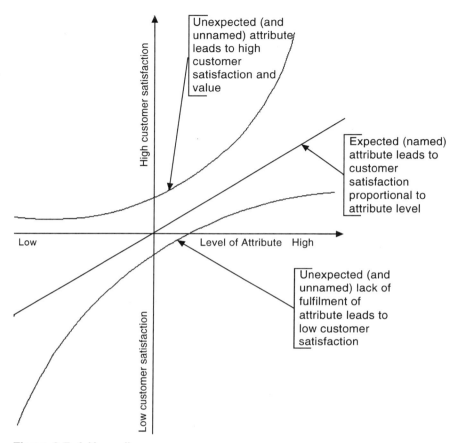

Figure 3-7 A Kano diagram.

decision-analysis-based financial models for all potential opportunities, using a consistent methodology, then one can conduct a portfolio analysis. Typically, there are constraints beyond the goal to provide the most NPV for a selection of projects. These constraints might be for specific revenues and costs, each required on a yearly basis. Linear programming techniques are often used to solve these constrained optimization problems. An example of the type of analysis provided by portfolio analysis is shown in Figure 3-8, which ranks the cumulative amount of profit against the cumulative cost each project will contribute.

A Caveat About Using Decision Analysis

Decision-analysis-based financial models must be considered as aids for decision makers; they are not intended to be followed blindly. These models

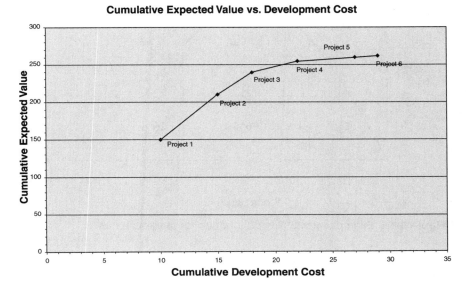

Figure 3-8 Example of a portfolio analysis.

supplement judgment and intuition. Good outcomes don't always result from good decisions, nor do bad outcomes always result from bad decisions. Thus, if one wins the lottery, this event is a good outcome (winning) from a bad decision (buying tickets).

REFERENCES

1. Raiffa H. Decision analysis. Reading, MA: Addison-Wesley, 1968.
2. Howard RA and Matheson J, eds. The principles and applications of decision analysis (2 vols). Palo Alto, CA: Strategic Decisions Group, 1983.
3. Matheson D and Matheson J. The smart organization. Creating value through strategic R&D. Boston, MA: Harvard Business School Press, 1998.
4. *http://www.real-options.com/.* Accessed 4 April 2002.

Appendix
How Expected NPVs are Calculated

To show how expected NPVs are calculated, let's return to the launch date example. We have three probabilities and their corresponding market shares, as shown in the following table.

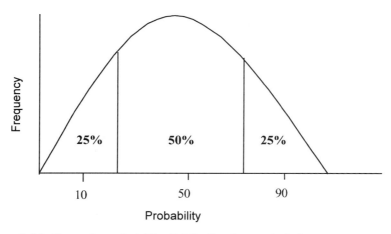

Figure 3-1A Example probability distribution for market shares.

Intuitively, we might try to multiply the probability column by the market share column. However, the column of probabilities doesn't add to 1, so this doesn't make sense. What is really happening is shown in Figure 3-1A. The x-axis numbers represent the 10-50-90 inputs. The bold numbers are the associated probabilities for these inputs.

The 20-month launch date in Table 3-1A is not really associated with 0.10 probability, but covers all probabilities from 0 to 0.25. The 0.10 number is a facilitator tactic to elicit a proper response because it is easier to ask for a 1-in-10 chance than to try to explain that this number represents all probabilities from 0 to 0.25. Similarly, the 0.50 number represents the probability range from greater than 0.25 to 0.75, and the last 0.10 represents the probability range from greater than 0.75 to 1.0. Now, we can calculate the weighted market share percentage = $(0.25 \times 30) + (0.50 \times 25) + (0.25 \times 15)$ = 23.75 percent. Thus, according to this example, the expected market share is 23.75, which differs from the most likely or base case market share of 25 percent.

Table 3-1A
Launch Date Probabilities and Market Shares

10-50-90 case	Probability	Launch date	Market share (%)
10	0.10	20	30
50	0.50	30	25
90	0.10	60	15

4

Stage II: Proving Feasibility

But it's what the customer wants! A last-ditch plea by an R&D scientist to keep his or her feature in the product specifications.

SETTING PERFORMANCE SPECIFICATIONS USING QUANTITATIVE METHODS

The Importance of Adequate Performance Specifications

As discussed in Chapter 3, the way the diagnostic assay business is managed and the decisions made can mean a lot to R&D scientists and engineers. If a project is no longer perceived to be financially rewarding, often because it is behind schedule, it might be canceled. Adequate specifications (the word specifications is used interchangeably with goals and requirements) can help prevent projects from being late. Specifications provide a stake in the ground and can provide a signal for a project manager to stop development when certain goals are reached. Otherwise, a scientist might continue to perfect an assay at the expense of its being late. Moreover, a specification with adequate evaluation criteria speeds decision making because it allows a project manager to determine more quickly whether the specification has or has not been met.

Figure 4-1 shows how specifications relate to the entire evaluation process and why the product-development process is so challenging. The items that present difficulties are inside the irregular shapes in Figure 4-1.

At the start of a project, one must select, from the large universe of possible goals, a set of specifications that become the product's design requirements. Marketing research tools help to correctly identify these goals. To determine whether a goal has been met, one must perform an evaluation based on an experimental design that attempts to allow one to observe the true state of nature, which is not directly observable. Proper data analysis and reports extract the information contained in the raw results from an evaluation.

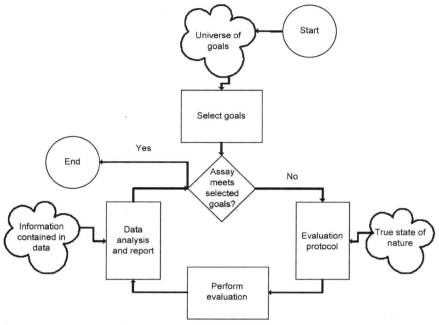

Figure 4-1 Process of evaluation in the product development cycle.

Adequate and Less-than-adequate Performance Specifications

Surprisingly, performance specifications are often inadequate, despite their importance. Preparing good performance specifications is difficult, in part because they describe goals for parameters that are estimated with uncertainty. Before describing the elements of a good performance specification, let's enumerate the several types of inadequate specifications.

Nonexistent Specifications

When no specification exists, many things are simplified. No protocols need to be developed or analyses conducted, and no release/no release decisions needs to be made. However, the lack of a specification does not mean that no problems will occur; it means only that the limit at which problems will start appearing is unknown. The allowable rate of performance outliers is an example of a performance characteristic that is often not specified. Yet products have been recalled from the market place for failing this implied performance specification. Product recalls, initiated either by the manufacturer or by the FDA, are listed on the FDA Web site (*1*).

Nonquantitative Specifications

A typical performance parameter, such as imprecision, might initially be specified "as good as or better than the competition" or "good enough to not cause complaints." If not quantified, these ideas will be difficult to evaluate.

Some performance parameters may be left as qualitative because it might seem as if quantification is not possible. For example, today's instrument systems are often specified to have a high degree of "ease of use." Yet, one can perform experiments to quantify or compare ease of use, which means that it is possible to create a quantitative specification. The following is a brief example protocol that demonstrates quantifying ease of use.

To simulate customer instrument operators, one selects participants with a desired background, typically with no prior knowledge of the instrument under test. Each participant is provided with a specific amount of training and asked to perform a series of tasks. An observer keeps track of the mistakes made, if any, and the number of times the test user consults a manual or asks for help. To conclude this type of experiment, the user is surveyed about his or her experience. The survey provides a ***perception*** of the user's experience while a tally of the mistakes, and times and types of help sought provide the quantitative measure of ***performance*** that was achieved for ease of use. Quite commonly, the perception does not match the performance.

Unrealistic Specifications

As an example of an unrealistic product specification, a product manager tried to set a reliability specification for a new analyzer at four unscheduled service calls per instrument per year. This new analyzer was the most complex instrument of its kind yet to be built. Analysis of similar products showed that their service call rates were greater than 10. Although anything is possible and "stretch" goals at times can motivate people, there were no new techniques or resources proposed for the new analyzer that would suggest that this goal was attainable. In the end, this type of goal becomes ignored by product design engineers and in that sense is similar to a nonexistent specification.

Software error rates often seem to defy being correctly specified in the diagnostic industry and could be considered to fall into this category of unrealistic specifications. One often hears, "We will not release the product until there are no serious software bugs." One must actually specify software quality as a rate—namely, the software failure intensity (2, 3). This rate provides the number of allowable failures observed within some time measure (CPU cycles or calendar time). For complex software, which includes most software in diagnostic instrument systems, it is impossible to test all combinations of software code—which makes it impossible to prove a failure intensity of zero—because multiple software branches yield too many combinations to test. This, coupled with the fact that software failures are found throughout development, makes the goal of zero software failure intensity unrealistic, even when one restricts failures to serious events. The situation perpetuates itself because the results of actual software testing programs are often stated in a misleading way to try to show that the goal of no software bugs has been met. They include statements such as

- All software modules have been tested successfully.
- We have fixed all bugs.
- We tested the software as a customer for a specified length of time and found no bugs.

Ironically, it is entirely possible that the software quality and associated testing programs will be adequate in spite of the software quality being incorrectly specified, the required metrics not being assessed, and the actual software failure intensity being greater than zero.

Incorrect Specifications

Sometimes specifications are incorrect. This is often simply a case of someone not understanding how to correctly formulate a specification. For example, an interference goal might be stated as "no interference from aspirin." This goal is impossible to verify. What is really meant is that the interference from aspirin must be less than 1 mg/dL — an amount that will not cause problems and can be verified. In another case, a project manager requested that the CV at the detection limit be 10 percent or less. This is impossible because, by definition, the CV at the detection limit is very high (4). The NCEP (National Cholesterol Education Program) proposed limits for total analytical error that are based on the assumption that total analytical error is equal to the mean bias + 1.96 times the total CV. (5). Yet this method neglects the effects of random biases (6).

Specifications Without an Associated Testing and Analysis Method

Assume that we have a clearly stated quantitative specification such as "the imprecision for the 50–150 mg/dL glucose range should be 2 percent CV or less." If no evaluation protocol is provided, how does one verify whether this specification has been met? Do we test 10 samples or 1,000 samples? Over one day or one month? Do we use a point estimate or a confidence interval? If the latter, at what probability level? Can we throw out a "bad" value or repeat a suspect run? What happens to the testing method being used if the project manager leaves in the middle of the product-development process and is replaced by someone else? These questions demonstrate why a proper specification requires both a verification protocol and an analysis method. The protocol describes the sampling method and conditions to be used, and the analysis method defines the statistical analysis and reporting method to be used.

Characteristics of an Adequate Performance Specification

An adequate performance specification will

- Be quantitative
- Be understandable
- Be measurable and ideally allow future values to be forecast
- Contain evaluation criteria to provide a means to determine whether the specification has been met

An Example of a Performance Specification for Blood Gas Analyzer Glucose Imprecision

The within-run CV for glucose should be 2 percent or less for the range 50–150 mg/dL. The protocol that should be used is to assay 10 consecutive samples as described in the NCCLS protocol EP10 for each of five runs. This protocol uses three concentration levels, which in this case should be 50, 100, and 150 mg/dL. The within-run CV is calculated as described in this guideline.

This specification meets the criteria suggested above and can also be used to predict future values, although that aspect is not contained in the specification.

How Specifications Change Through the Development Process

Setting product specifications often follows the scenario depicted in Figure 4-2. When the project starts, people are genuinely excited. The product concept is interesting, the sales and profit projections are high, and the sense is that this will be the best product the company has ever produced. These expectations are transferred into lofty performance specifications. Yet during feasibility, as some goals are not met, the rationale for some of these goals is reexamined, and perhaps some goals are lowered. Finally, when it is close to release for sale, and if goals are still not met, this examination recurs. It is often during the release-for-sale meetings that the best discussions about performance specifications occur! An improvement would be to transfer the quality of this meeting back to the project start.

Different Origins of Performance Specifications

Specifications for the same performance parameter, such as imprecision, often come from at least three different sources (7). The most stringent of the three

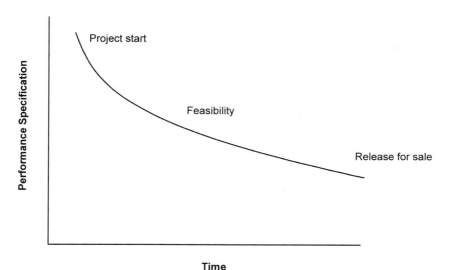

Time

Figure 4-2 Performance specification goals decrease as a project progresses through the product-development cycle.

specifications becomes the specification that must be met. The common sources are regulatory, medical need, and competitive.

Regulatory

Real or perceived regulatory specifications are those performance requirements required for approval by the FDA in the United States or by similar agencies in other countries. In addition, one has to take into account quality control and proficiency survey requirements to ensure that the assay will satisfy these after commercialization.

Medical Need

Medical need goals are the set of performance requirements that will ensure that assay result error is low enough to prevent incorrect medical decisions because of laboratory error. Agreement has been difficult to reach on these goals, which are developed by different methods (8, 9). Companies often have a physician on staff to address setting these goals.

Competitive

These are usually the tightest specifications, provided by marketing managers, and are needed (or perceived to be needed) to gain or maintain market share. Because these specifications are the most stringent, they are often relaxed if there is trouble in meeting them. They can be relaxed until they run into the regulatory or the medical need specification level.

In addition to these factors, new information, such as the release of a competitor assay, new therapies, or a variety of other information, might cause specifications to change.

How Specifications Are Used Differently by Manufacturers and Customers

Manufacturers are faced with the decision to release or not to release (and therefore scrap) products. Specifications simplify this decision to "Does the product meet specifications?" Yet consider Figure 4-3, which shows a set of different reagent lots and their associated accept/reject criteria for the specification bias (*10*). The curve represents a conceptual distribution of biases due to many reagent lots and indicates that most reagent lots produced will fall with the specification limits. Considering the lettered example lots in the figure, from a manufacturer's financial perspective, lots A through C have equal and full value and will all be released. Lot D is out of specification, will be scrapped, and is considered to have zero value. The financial value of lots to the manufacturer is represented by the dashed line.

Consider the customer's perspective and assume that by some means Lot D has escaped from the manufacturer and is in the customer's hands. Unlike the manufacturer's view, Lots C and D have similar performance and must therefore have similar value. The conceptual financial value to the customer is identical to the solid curve representing the bias distribution. Its value is maximum at zero bias and falls away in a quadratic fashion as bias increases.

Note also that Lots B and C, both of which conform to the manufacturer's specifications, could actually cause problems for a customer if these two lots

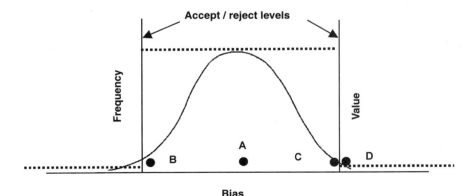

Figure 4-3 Performance implications for various reagent lots.

were in the lab at the same time or nearly the same time. Moreover, should the customer complain, they might be told that there is nothing wrong with either lot, which is true from a manufacturer's perspective. The fallacy in the manufacturer's reasoning is that lots are considered acceptable for the amount of bias they exhibit compared to zero bias (lot B minus A). But from a customer's perspective, the amount of bias under consideration is double the manufacturer's amount (lot B minus C).

One financial solution to these quality issues is to value each lot differently (e.g., have a selling price that reflects the lot's closeness to zero bias as shown by the customer financial value curve). However, this solution would be currently unacceptable to many diagnostic customers and manufacturers. The best solution is to shrink the distribution of observed bias in reagent lots, so that no lot or combination of lots will cause a customer complaint and also that the number of rejected lots will be negligible. This is the principle of continuous quality improvement.

Specific Techniques Used to Set Performance Specifications

Whereas it would seem easy to obtain correct performance specifications by simply asking potential end users, this is actually a more difficult practice than one might imagine. The difficulty is twofold:

- Manufacturers have trouble asking customers what they want
- Customers have trouble describing what they want

The following section expands on this and provides techniques to obtain better specifications.

Focus Groups and Surveys

Focus groups represent a structured way for a manufacturer to talk to a group of customers to glean what they want and are willing to pay for in a product. Oftentimes the identity of the company is withheld, although this may be difficult if a prototype product is displayed in the focus group. A facilitator plays a key role to ensure that there is the right balance of participation from the attendees. In addition, the facilitator must ensure that he or she does not bias the answers by ensuring that the questions asked are neutral. The output of a focus group is rarely quantitative but forms a basis for selecting product attributes that might require quantification.

Surveys attempt to obtain this information by mail, by phone, or in person through interviews. There is always the danger that surveys will be completed by someone who has not understood the questions. Another issue with surveys is that customers normally provide responses that request the best of everything: an assay that has outstanding performance, low cost, high reliability, high ease of use, and so on. In real life, one must make tradeoffs, which can be assessed through methods such as conjoint analysis.

Conjoint Analysis

Conjoint analysis is used to study factors that influence a customer's purchase decision, especially when the customer is faced to make tradeoffs among the various factors (*11, 12*). Consider the following problem of determining which features on a new routine chemistry analyzer are most highly valued by customers.

Analyzer Features

Price	$30,000, $40,000, $50,000
Menu (tests offered)	50, 40, 20
Stat time to result (min)	2, 5, 10
Onboard ISEs	Yes, No
Design shape	Stylish, Boxy

If a manufacturer conducts a survey to answer the questions above, the rational response is for everyone to want the best. In conjoint analysis, one asks potential users to rank each of several analyzers with sets of features selected from the above list. To present all possible 108 combinations ($3 \times 3 \times 3 \times 2 \times 2$) would overwhelm the user with too many choices. Therefore, subsets of choices are selected based on factorial design principles. Orthogonal arrays are often used as a minimum set of choices that will allow uncorrelated estimates of main effects. Continuing with the analyzer feature problem, an example of the subset of the 108 possible analyzers is shown in Table 4-1. A set of cards, randomized from the list in Table 4-1, would be presented for the user to rank.

The data is then analyzed statistically to derive the value of each attribute. Worked examples using PROC TRANSREG in SAS are available on the SAS Web site (*13*). Tom Novak has an interactive Web site that allows one to go through a procedure interactively and see the typical output produced by conjoint analysis—a series of part-worth graphs that show part-worth vs. attribute level (*14*). Conjoint analysis results can be used for market share simulators and updated throughout the product development cycle as new data becomes available.

To see how conjoint analysis can save money and speed up product development, let's add some personalities to the feature selection problem. Consider an industrial design manager who insists on developing a complex yet stylish shape for an instrument, which will likely require other departments to tailor their designs to meet the constraints of the complex shape. This has all of the earmarks of delaying the project and causing additional expense. Yet the industrial designer may persuade management that his or her

Table 4-1
An Example of Analyzers to Be Ranked, After Randomizing This List

Number	Menu	Stat TAT	Price (in thousands of dollars)	Onboard ISEs	Design shape
1	50	2	30	Yes	Stylish
2	50	5	40	No	Boxy
3	50	10	50	Yes	Stylish
4	40	2	40	Yes	Stylish
5	40	5	50	Yes	Boxy
6	40	10	30	No	Stylish
7	20	2	50	Yes	Stylish
8	20	5	30	No	Stylish
9	20	10	40	Yes	Boxy
10	50	2	50	No	Boxy
11	50	5	30	Yes	Stylish
12	50	10	40	Yes	Stylish
13	40	2	30	Yes	Boxy
14	40	5	40	Yes	Stylish
15	40	10	50	No	Stylish
16	20	2	40	Yes	Stylish
17	20	5	50	Yes	Stylish
18	20	10	30	Yes	Boxy

design, which may in fact win design awards, will greatly enhance sales. The value of conjoint analysis is that it helps senior management make a decision based on data and not just on the ability of various project managers to argue their case. In this example, management could consult Figure 4-4, which shows two of the five part-worth graphs. The bottom graph provides evidence that the design shape has no value to customers. That is, customers are completely ignoring this feature when making their choice among different analyzers. This implies that the project delays and additional expense caused by the stylish shape would not be offset by additional revenue.

Quality Function Deployment

Quality function deployment (QFD) is a method that translates desires from the customer to the manufacturer. Figure 4-5 shows how these customer wants, which become the product specifications, are transferred throughout the entire product development cycle (*15, 16*). This is a key feature of QFD because even if the customer requirements have been correctly identified, many choices must be made as part of the overall product design that are only indirectly linked to

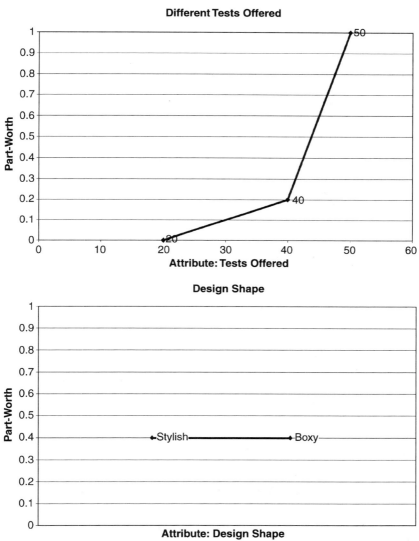

Figure 4-4 Part-worth graphs for two design attributes.

the original customer requirements. As in the conjoint analysis example, sometimes design choices may be based on reasons not related to customer preference. QFD helps management hear the voice of the customer rather than the voice of the manager.

The following outlines the QFD steps for the product planning block in Figure 4-5. Other blocks are carried out in a similar fashion. Central to the

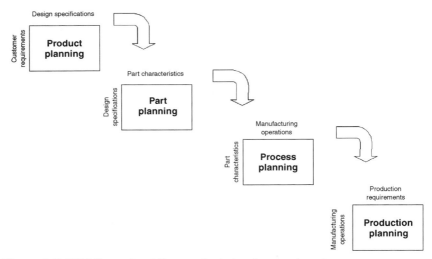

Figure 4-5 QFD throughout the product-development cycle.

success of QFD is a team comprised of marketing, R&D, and manufacturing people.

Identify and rank customer needs—Marketing studies such as focus groups, surveys, and conjoint analysis are used for this purpose. Customer needs are ranked in order of preference, which provides a better model of reality than "all customer demands are of equal importance," which one hears at times. The ranks are often created so that the total equals 100 percent.

Compare customer perceptions for competitive products—This benchmarking allows one to identify gaps relative to each product attribute between the company's existing products and its competitors, as perceived by customers.

Identify and correlate product design features—Product design features are the technical translations of the often-qualitative customer needs. One establishes correlations between each product design feature and customer need.

Quantify metrics for product design features—Comparisons are made for the company's existing products and its competitors for the metric corresponding to each product design feature. This is similar to the comparison of customer attributes.

Estimate risks and rewards—The imputed importance of each design feature, its technical difficulty, and cost are all estimated.

Decide on product design targets—Based on the above tradeoffs, targets are issued for each product design feature.

DIFFERENT APPROACHES TO DEMONSTRATING FEASIBILITY

Given that specifications are in place and preliminary designs have been constructed, one would like to know whether feasibility has been met. The demonstration of feasibility is simple in concept but elusive to define in practice. If one
has achieved the set of performance goals during a feasibility evaluation, then
why not immediately release the product? Alternatively, if performance goals
have not been reached, delaying the next development stage may cause the overall project schedule to slip unnecessarily.

Feasibility used to be easier to conceptualize. In the past, assays were often
developed for feasibility using "research lot" reagents specially made by R&D
scientists rather than by manufacturing. If performance goals were reached, the
assay was transferred to manufacturing, whose task was to make a reagent lot
that would pass performance goals. Today this distinction is often blurred.
Assays that require chip technology, for example, are developed within the manufacturing process, even for feasibility evaluations. The only difference between
a feasibility lot and a release-for-sale lot is the size.

In any case, one must decide what to do with the data collected during a feasibility evaluation. This is facilitated by understanding which question is being
asked. One conceptually thinks of feasibility as an answer to the question "Can it
be done?" Whereas this is sometimes the appropriate question, for many assays a
more common question is "When will it be done?" In both cases, the data show
that the assay has either met the feasibility goals or has not met them. If feasibility
has not been met, the most important question now becomes "Is the plan to meet
goals credible?" The plan may contain phrases such as "We need to use a more
sensitive amplifier," "We didn't add enough enzyme," and "We have preliminary
evidence that Triton X-100 will eliminate the bubble problem." The ability of management to evaluate these plans plays a key role in efficient product development.

Beware the Technical Administrator

For a decision maker to answer the questions posed above, a sufficient level of
technical understanding is required. The term technical administrator is used for
managers who are not technically qualified to answer these questions. Technical
administrators are commonly found in organizations. Some industry examples of
people who fit this description are the following:

- A vice president of R&D reports to the president of a company. The president
  is trained in business, not science and thus serves as a technical administrator
  for R&D
- A vice president of R&D is trained as a biochemist. Yet he manages all of
  R&D. His competence in engineering is marginal, so he acts as a technical
  administrator for the engineering aspects of R&D

- A manager heads a department for which he or she is not qualified technically. We won't enumerate the possible reasons that they are in their positions.

The first two instances of a technical administrator described above are not pejorative whereas the third instance is. In any of the above cases, the lack of technical understanding can cause problems in evaluating feasibility and for many other situations. The authority to make a decision does not mean that the decision will be the correct one or even that the decision maker knows enough to explore the issue intelligently. Technical administrators can decide on technical issues by any of the following means:

1. Ignoring the technical issues and basing the decision on other grounds such as relationships.
2. Letting the lead technical person make the decision alone. This is the unsupervised approach.
3. Convene one or more technical experts to recommend a decision.

Often the third choice is the most appropriate. As in the financial modeling case, providing written justification for decisions can improve decision making and can allow for continuous quality improvement in decision making.

REFERENCES

1. *http://www.fda.gov/opacom/Enforce.html*. Accessed 4 April 2002.
2. Musa JD, Iannino A, and Okumoto K. Software reliability: Measurement, prediction, application. New York: McGraw Hill, 1987.
3. Farr W. A survey of software reliability modeling and estimation, NSWC TR 82-171, Naval Surface Weapons Center, Dahlgren, VA, September 1983.
4. Krouwer JS. The CV at the detection limit. Clin Chem 1989;35:901.
5. Recommendations regarding public screening for measuring blood cholesterol National Heart, Lung, and Blood Institute, National Institutes of Health NIH Publication No. 95-3045 September 1995.
6. Estimation of total analytical error for clinical laboratory methods; Proposed guideline NCCLS EP21P. NCCLS, 771 E. Lancaster Ave., Villanova, PA, 2002.
7. Krouwer JS. "Use of analytical goals by health care manufacturers." In Proceeding of 1995 Institute on Critical Issues in Health Laboratory Practice. Frontiers in Laboratory Practice Research US Dept. of Health and Human Services, 1996.
8. Fraser CG. Biological variation: From principles to practice. Washington, DC: AACC, 2001.
9. Boone DJ. Governmental perspectives on evaluating laboratory performance. Clin Chem 1993;39:1461–1465.
10. Ross PJ. Taguchi techniques for quality engineering. New York: McGraw-Hill, 1988.

11. Green PE and Rao V. Conjoint measurement for quantifying judgmental data. J Market Res 1971;8:355–363.
12. Green PE and Wind Y. New ways to measure consumers' judgments. Harvard Bus Rev 1975;53:107–117.
13. *http://ftp.sas.com/techsup/download/technote/ts622.pdf* Accessed 27 April 2002.
14. *http://www2000.ogsm.vanderbilt.edu/novak/conjoint/* Accessed 27 April 2002.
15. Hauser JR and Clausing D. The house of quality. Harvard Bus Rev 1988;66:63–73.
16. Hauser JR. How Puritan-Bennett used the house of quality. Sloan Mgmt Rev 1993;34:61–70.

5

Stage III: Scheduled Development

I feel really good about this project. — A common project manager's summary statement at a project review

WHY PRODUCTS ARE ALMOST ALWAYS LATE

As described in Chapter 3, an early launch date is a high priority because it almost always provides improved financial performance. This creates management pressure on R&D to have the product ready sooner. R&D managers typically don't advance in their career by saying no when asked whether they can meet an aggressive date. Of course, in some cases, they will not be allowed to say no even if they wanted to but will be told, "Find a way to make it happen." Most scientists and engineers accept this challenge. Oftentimes the result is agreement with an overly optimistic schedule followed by a missed launch date.

A project that is late, which is tantamount to a broken promise, often prompts management to search for solutions to make R&D more efficient. Yet, a late project may simply be attributable to a scheduling problem. One way to provide insight into this problem and to improve schedule accuracy is to apply the 10-50-90 method described in Chapter 3 to product-development schedules, even if one does not otherwise use decision-analysis–based financial models. Thus, rather than having only one launch date, one would have three launch dates, each with its likelihood of occurrence: an optimistic (1 in 10 chance), most likely, and pessimistic date (also 1 in 10 chance).

If only a single launch date is used (see Chapter 3 for the problems associated with this), one can still analyze that date in probability terms. That is, an overly optimistic launch date will be achieved, only if everything in the project goes perfectly (e.g., each task will have an optimistically short completion time). The sum of these tasks will equal the overall optimistic date, which is theoretically possible but very close to having a probability of zero.

Figure 5-1 shows the cumulative probability for a conceptual series of nearly identical projects that range from 30 to 70 months in duration. A probability close to zero means that if many similar projects were carried out, almost none of them would be completed by the optimistic 30-month date, which in turn is a

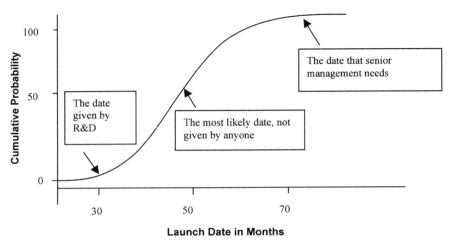

Launch Date in Months

Figure 5-1 Probabilities for launch dates provided by different groups.

manifestation that it is extremely rare that everything goes perfectly on each task within a project. The most likely date means that half of the projects will be released later, and half earlier than the 50-month completion date. Finally, the 70-month date means that virtually all projects will have been completed by this date.

It is ironic that marketing and commercial managers actually need to know the launch date that has a cumulative probability close to one (e.g., the date farthest away from that given by R&D). This is because these managers need to ensure that training, advertising, and other launch activities are not held too early. Also, commercial managers often discount the currently sold product shortly before the new product is released. This must be done at the right time for optimal financial results.

One technique that senior management employs to take into account Figure 5-1 is to use rules of thumb to add time to launch dates provided by R&D managers. Senior managers rationalize that R&D is always late, and thus simply account for this by applying some multiple to the proposed schedule time.

Other, more data-driven techniques can also help management diagnose projects that might be later than their promised dates. For example, Silverberg (1) presented a method that is illustrated in Figure 5-2. Here, predicted completion dates (y-axis) are plotted against the dates the predictions were made (x-axis). The identity line is shown on the graph as the "completion line." An ideal project would appear on this graph as a horizontal line that intersects the completion line. A worst-case project would appear as a horizontal line that is just short of intersecting the completion line and then veers parallel to the completion line. In management terms, this is a last-minute surprise, which is almost always considered the worst possible scenario. As an illustration of Silverberg's method, consider the two projects shown in Figure 5-2.

Data Driven Schedule Predictions

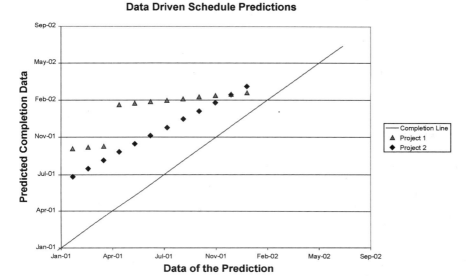

Figure 5-2 Using past predictions to assess schedules.

Project 1 is predicted to be completed by October 2001, with the first prediction made in January 2001. Each month, a new prediction is made that changes little until April 2001, when a rather large schedule slip emerges. Thereafter, the predictions continue to hold mainly to the new date of February 2002.

Project 2 is predicted to be completed by July 2001, with the first prediction also made in January 2001. Each month, the new prediction made for Project 2 shows a small but consistent schedule slip. The last prediction calls for the project to be completed by March 2002, very close to the predicted completion date for Project 1.

In the example, the date of the most current prediction for both projects is January 2002. With the ability to see the patterns of the previous predictions, it is likely that Project 2 will be completed long after Project 1, based on past predictions. Unfortunately, most project scheduling software does not maintain this type of data, but overwrites historical predictions when schedules are updated.

Another technique to assess schedules begins with the observation that R&D projects almost always have unanticipated problems. It is the failure to account for the time required to solve these problems that is responsible for incorrect schedules (2). Psychologically, one can see the difficulty that an R&D manager might have in trying to add this additional time to schedules, which are usually already under pressure (i.e., "Do we really need that much time for activity XYZ"? What R&D manager would care to admit that he or she would like to add six more months for a series of unanticipated problems that might occur?)

The situation is made worse by using misleading metrics. For example, the metric "percent completion of tasks" has been used as a measure of progress. The problem with this metric is seen in the following example. Management is told that in five months, 50 percent of the tasks have been completed, with the implication that the launch date will occur in another five months. But this will be true only if task-completion times are uniformly distributed. Yet often, the easy tasks are completed quickly, with the harder ones taking longer than expected. This is equivalent to an exponential distribution of task-completion times. The result is that the project still may be a long way from completion even when 95 percent of the tasks have been completed.

The false sense of near completion is made worse with the use of the currently popular dashboard measures (3). These are graphical displays similar to a car's dashboard instruments that are used to display progress. However, most people expect a proportional linear relationship between the input variable (time) and the output variable (percent completion) in such a display.

One can try to model historical task completion times (2) to be able to predict launch dates. Alternatively, if one knows what the longest task will be, then one can devote modeling efforts solely to estimating the duration of that task. For example, in large, complex instruments systems, reliability is often the gating item for launch. Reliability growth management methods (discussed later in this chapter) have been shown to be useful in predicting the duration of this activity.

The preparation of a typical "bottoms-up" schedule whereby each activity's duration is predicted, and task overlaps and dependencies are accounted for is an accepted and necessary practice. Use of the above techniques is recommended to supplement the bottoms-up method.

USING DESIGN OF EXPERIMENTS METHODS TO BUILD ROBUST ASSAYS

A robust assay is one that always meets its performance specifications regardless of manufacturing tolerances and variation in environmental variables, both of which are inevitable. Design of experiments (DOE) methods help increase the likelihood that assays will be robust and can shorten development time compared to non-DOE development techniques. DOE methods are appropriate when the state of knowledge is medium (see Chapter 3).

Why Many Scientists Do not Use Design of Experiments (DOE) Methods

Some scientist and engineers are reluctant to use DOE methods because DOE was not part of their training and also because many scientists and engineers have been quite successful without the use of DOE. In addition, DOE can initially require involvement of a consultant. This may be threatening to the scientist or engineer because it represents a potential loss of control. DOE methods must be positioned correctly in an organization if they are to be accepted. They do not replace creativity and are not even always appropriate (for example for the case of a high state of knowledge). Yet in many cases, they can help develop robust products more quickly.

Cause-and-effect Diagrams and Process Flow Charts

The starting point for DOE methods is to acquire knowledge of the influential variables. Typically, there will already be considerable knowledge of these variables. In some cases, where there is very little knowledge of variables, screening experiments or perhaps one-at-a-time experiments may need to be performed. Slater (*4*) calls the influential variables key input variables (KIVs), and the important output variables the key output variables (KOVs). The KOVs are known collectively as quality and are totally dependent on the KIVs.

Knowledge of the relationships between key output and input variables can be formalized with cause-and-effect diagrams and process flow charts. A cause-and-effect diagram is often conducted as part of a brainstorming session. As with the data input part of financial modeling discussed in Chapter 3, a facilitator plays a key role in the brainstorming session for a cause-and-effect diagram. Brainstorming is a *managed* free-for-all exchange of ideas. Without managing the brainstorming session, it is possible to get bogged down in minutiae and not produce useful output. The types of cause-and-effect diagrams follow, with the definitions somewhat arbitrary. For instance, a fishbone diagram is explained as a type of cause-and-effect diagram, but a cause-and-effect diagram could be explained as a type of fishbone diagram.

Cause and effect diagram — A cause-and-effect diagram (see Figure 5-3) contains an undesirable top-level event. Each box is a cause for the effect above it.

Fishbone (Ishikawa) diagram — A fishbone diagram is a cause-and-effect diagram turned on its side (rotated 90 degrees clockwise).

Fault tree — A fault tree is a cause-and-effect diagram often used by engineers for systems. Some fault trees quantify the probability of top-level or other events through algorithms for combining probabilities. Since fault trees are constructed starting with a top-level event and building downward, they are sometimes referred to as "top-down systems." Fault trees and Failure Mode Effect and Criticality Analyses (FMECAs) are often created at the same time. In a fault tree, two or more events are connected to an above event through a logical operator. The two most common operators — or gates as they are called — are as follows:

- An **AND** gate means that the top-level event occurs if *all* events connected through the gate to the top-level event occur.
- An **OR** gate means that top-level event occurs if *any* event connected through the gate to the top-level event occurs.

When an event has no more events connected below it, it is called a basic event. FMECAs are always prepared for basic events.

The use of fault trees is not restricted to _____ example, linear drift might _____ _____ formance. The parent event _____ an **AND** gate to one event _____ he speci- men assayed since _____

Failure Mode _____ CAs con- tain more details _____). Since FMECAs are con _____ they are sometimes referre _____ ned in a FMECA (here the _____

- *Cause* — The roo _____ or un- known.
- *Effect* — The resu _____ n.
- *Fault detection* — _____ mer, or the service depart _____
- *Fault isolation* — Given that the failure is detected, the means through which the failure is isolated to a specific component.
- *Fault recovery* — The procedure that returns the system to an operational state.
- *Criticality* — The importance of the event, such as patient safety, unscheduled service call, customer recoverable, or poor performance.
- *Probability of occurrence* — A mapping between words, such as frequent or extremely unlikely, and their approximate rates, such as "probability of occurring once per year is 1 percent or less."
- *Action taken* — Identifies what was done to minimize the problem, such as a design change, additional built-in test (BIT) deployed, or manual change.

Hazard analysis — Hazard analysis is a fault tree, often required by regulatory agencies, which focuses on patient and operator safety. It often has branches for software faults, environmental problems and operator safety issues, and incorrect assay results.

Factorial and Response-surface Methods

Given that the key influential variables have been identified, factorial methods represent a means of determining optimal settings for these variables. Factorial methods can be divided into the following categories:

Screening designs — These experiments are a type of fractional factorial design used to determine which of several possible variables (5 to 11 variables are typically screened) most influence the output variable under study.

Fractional factorial designs — These experiments are the most common factorial designs used to optimize assay responses. That is, levels of factors are found that maximize signal or minimize imprecision. These designs involve

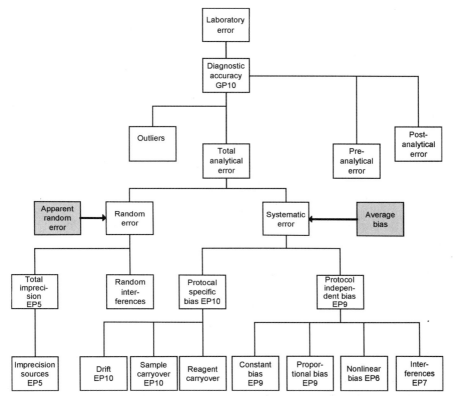

Figure 5-3 Example of a cause-and-effect diagram. EP and GP numbers refer to NCCLS protocols.

testing two levels of variables, usually numbering 2 to 6. The designs are fractional when not all possible combinations are tested. For example a 2^{4th} factorial design involves 16 separate runs, which includes all possible combinations of two levels of four variables. A $^1/_2$ fraction is called a 2^{4-1} design and is performed with eight runs. It can be shown that for k variables, a one-at-time design requires $(k + 1)/2$ times as many runs as that required for a factorial design (6).

Response-surface methods—These are factorial designs that include three or more levels of at least some of the variables (often 2–4). With three levels, quadratic effects can be estimated (curvature), which allows for more detailed response-surface maps.

There are many excellent descriptions on the theory and practice of design of experiments. Perhaps the best book is that by Box, Hunter, and Hunter (6). The rest of this section focuses on using a experiment-planning checklist, a practical implementation aid that can help in the use of these factorial designs when a statistician is not available.

Experiment-planning Checklist

Use of an experiment-planning checklist helps with all experiments, not just with factorial designs (7). Its use is intended to make sure that important steps in an experiment are considered. This is ensured by having the experimenter record key steps that are planned before the experiment is performed. The benefit of this approach is that it requires one to think of experiment details that otherwise might be missed. Additionally, it presents an opportunity for development scientists and engineers to communicate with statisticians. Ideally, the process is fashioned such that it requires only a small amount of extra work and does not constrain thinking. Menu-driven software can facilitate the process. The elements of a general checklist are outlined below.

Demographics — Besides all relevant names, this section should include the expected start and completion dates of the experiment.

Purpose of experiment — This obvious step is sometimes a challenge for the experimenter to put down on paper. A clear purpose has objectives that are specific and measurable and typically address a specification. In addition, all parties should agree on the objectives, the criteria for determining whether they have been met, and to agree on changed objectives, if needed.

Background of the problem — A statement of the problem's background should include a trail to previous experiments, which may involve different approaches in solving the problem. Problem genealogies in the form of treelike diagrams that are enabled though software can be very useful when a problem arises that was attacked a few years ago, or even a few months ago. The treelike genealogy helps when no one remembers the trail of experiments that were performed.

Response variables — The output variable is usually easy for most scientists to describe, such as the assay response (with units provided) or the assay imprecision.

Control variables — The obvious control variables are those that are adjusted to different levels as part of the experiment design. However, not all control variables are varied. All influential control variables that are held constant should be listed, as well as the level used for the experiment. The noise or environmental variables, which either cannot be controlled or cannot be allowed to vary, should also be listed.

Possible interactions — Using control variables A and B as an example, an interaction occurs when the slope of the response going from low to high A is different when B is held low versus when B is held high. Interactions affect the choice of experimental design.

Experiment design chosen — The type of experiment chosen must match the constraints imposed by the variables listed as well as afford sufficient precision in quantifying the results.

The results of an experiment are contained in a report. Reducing time to market is not achieved by R&D breakthroughs alone. Good versus poor reports can also provide efficiencies. For example, if the information contained in a well-constructed report is acquired by a 30-minute reading versus requiring a 4-hour meeting—which itself needs 2 days to be scheduled—all to discuss a poorly written report, one has saved two days for this single event.

Writing Reports That Convert Data into Information

The Need for Written Reports

There is sometimes a tendency to forgo writing a report and instead to verbally summarize the results of a study, almost always to save time. This should be avoided. When one writes a report, the conclusions or recommendations often change from those in a verbal summary. When one issues a written report, a permanent record is associated with the report's author. On the other hand, verbal summaries are quickly forgotten, especially if poor minutes are taken or if no minutes are kept. The permanent record subtly influences the author to scrutinize conclusions and recommendations more closely than would be done for verbal summarizes; hence the possible changes.

Reports considered here are limited to those based on data analysis. The basis of a good report is that it transforms data into information: the report does the work, not the reader. In this context, data and information are defined as:

Data—Facts and figures

Information—Knowledge gained from data

As an extreme example of the difference between data and information, the text below Figure 5-4 is an alternative representation of the figure (the text is the uuencoded transformation of the figure's binary content).

To some readers, the text representation above is what some reports look like.

Tips for Converting Data into Information

Raw data should always be summarized in some way. Plots and tables should be presented, as should statistical analyses when they will provide insight into the data. Although these recommendations may seem obvious, it is surprising how many meetings are spent discussing each of many individual data points, primarily because of the lack of an effective data summary. Although brainstorming with the raw data may be beneficial and is very common, the raw data should not be considered as "the report," with brainstorming considered as a review of the report.

Some specific examples of report problems follow:

Since two analyzers are being compared, the average results from analyzers A and B are presented in two columns. What is missing here is a third column that contains the difference between the two analyzers. If this third column is not presented, readers will be forced to construct it.

```
begin 666 burst1.gif
M1TE&.#EA/``\`+,`+,+,``"\O_U-3_W%Q_Y"0_[*R_]/3__W]_____P`````````
M`````````````````````"`"`"`````~'Y!`$`$$``````.``<4(/```````.`
M97((@5F@V
M97(((IY!U[^~((Yh:!1~(1TE&(%-M87)T4V%V
M97((@5F5R(#(N, ``L````#P`/```!!!!/_`/____!_`!_!/!_PW_/`____
MK"#j-`/(n:B-%=&D,3'%=#$DY;_-(#_G'8.4&!&@N~GF+GL;!:-
..............
```

Figure 5-4 A picture (top) and an alternative representation (bottom).

Some scientists use terms and units they are used to working with, not acknowledging that most people will not understand these terms and units. For example, a sodium result of 154 mmol/L is more understandable to most people than a sodium result of 23 mVolts.

Some reports are written as one very long paragraph. Others are presented as a story, perhaps written properly scientifically, but requiring the reader to digest the entire report to get to the conclusions or recommendations. Some reports do not have recommendations at all.

A Suggested Report Format

A report format that highlights information, not data, follows.

Purpose describes why you are writing the report (that is, why was the experiment performed?).

Background contains introductory information about the project and often contains an outline of the protocol. The full protocol is often appended.

Recommendations are actions such as: *use* 1 mmol/L phosphate (rather than: 1 mmol/L was found to be optimum). Recommendations are based on conclusions.

Conclusions are a concise summary of the results section. Conclusions are based on results.

- Individual recommendations and conclusions should be numbered and placed in separate paragraphs

Results are a description of the assumptions, data analysis methods, theory, and so on. Results are based on data.

- Results contain data summaries, tables, and plots.
- This is also a good place to document the system configuration (i.e., instrument serial number, software revision, lot numbers).

Data are the numbers or inputs to the experiment. Data can also contain summaries, but a trail to the raw data should be in place.

Attributes of this report format are as follows:

- Going upward from the data to recommendations sections
  - Data are being transformed into information
  - Sections get shorter
  - For a correctly written report, each section (recommendations through data) is supported by the one below it.
- People know where the recommendations and conclusions are.
- People who only want to read the recommendations can do so quickly.

Symptoms for Problem Reports

Poor reports are often not read at all or are skimmed, although the skimming is also superficial. In these cases any of the following scenarios can occur:

- People call and ask "What's the bottom line?"
- You are asked to call a meeting to discuss the report.
- You get no response because no one has read the report.

Using the suggested format report can prevent some of these problems.

USING RELIABILITY GROWTH MANAGEMENT TO BUILD RELIABLE SYSTEMS

Much effort in the diagnostic industry is devoted to developing assays with good analytical performance. However, regardless of how good an assay is, if it is unavailable because of a system failure, the laboratory and the manufacturer will suffer in some way. Most systems are sold with at least a 1-year warranty that provides on-site service. A service call resulting from an instrument failure can cost around $1,000 per incident. Large systems often can have 10 or more unscheduled service calls per year for systems under warranty. If there are 1,000 systems in the field, this represents a cost of $10,000,000 to the manufacturer. Moreover, for complex systems, failure to meet a reliability target often delays product release, which causes a delay in revenue and prevents engineers from working on other projects. Thus, industry is motivated to reduce the time it takes during development to reach a reliability goal.

An Overview of Reliability Growth Management

Reliability is defined as the probability of a system or component to perform its required functions under stated conditions for a specified period of time (8).

Reliability growth is defined as the positive improvement in a reliability parameter over time due to changes in product design or the manufacturing process (9).

Reliability growth management is based on learning curve theory and is credited to Duane (10), who observed that reliability improvement for a variety of different systems followed a similar pattern. The improvement was proportional to the cumulative time the system was under test (see Figure 5-5).

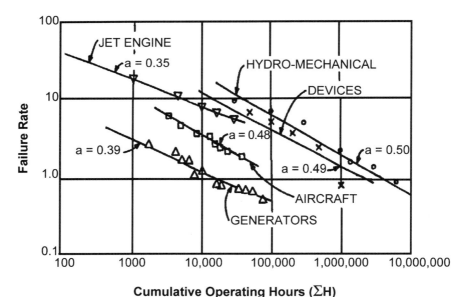

Cumulative Operating Hours (ΣH)

Figure 5-5 Reliability growth for different systems. (Reproduced with permission from Learning curve approach to reliability monitoring. IEEE Trans Aerospace. Copyright 1963 IEEE.)

These initial observations were developed into a technique that has become widely used in the defense and auto industries (9) and has been adapted by McLain into the diagnostic industry (11).

In reliability growth management, a system is put under test, which often simply means that an instrument system is operated to produce assay results. This provides an opportunity for the system to fail. Failures are observed and categorized. Fixes for the observed failures are proposed and applied. This cycle of testing, observing failures, fixing failures, and retesting is continued until the desired reliability goal is reached.

There is an important difference between reliability growth testing and reliability demonstration testing. The latter type of testing is performed *after* product design and manufacturing is complete and for a predetermined amount of time. In reliability growth testing, reliability is measured *during* the design phase, when the system is repeatedly being changed.

When "Testing in Quality" Is More Efficient Than "Designing in Quality"

Quality initiatives favor "designing in quality" or doing it right the first time, and take a dim view of "inspecting in quality" (12). Designing in quality is preferred because of the implicit assumption that it is more efficient than inspecting a product or process to determine what needs to be changed after it has been designed. In reliability growth management, "testing in quality" is a more accurate term, but this is still fairly close to "inspecting in quality," since one is fixing a product with defects. In fact, "designing in quality" and "testing in quality" are not mutually exclusive. Each should be used where appropriate, even on the same program. The choice depends on the state of knowledge of the technology (see also Chapter 3).

That is, when there is a high state of knowledge (e.g., when equations describe knowledge), then one can and should design in reliability and avoid testing in quality. However, for complex diagnostic instrument systems, the state of knowledge is moderate, and empirically based experiments are used to gain knowledge. There are often hundreds, if not thousands, of ways a complex system can fail. Furthermore, the means of mitigating these failures are often not known with certainty. With a moderate state of knowledge, reliability growth methods are often the fastest way to develop or improve products by testing to expose problems, proposing solutions, and retesting.

A Model of Instrument System Service Calls

The unscheduled service call rate for instruments under warranty (shortened here to "call rate") is usually the most important reliability parameter. A model of the call rate is needed because it allows one to measure and predict reliability for systems before they are released for sale. Traditionally, one thinks of reliability problems in terms of "hard failures." For example, if a power supply fails, the system shuts down and is unavailable until the power supply is replaced (usually by a service person). However, based on examination of service records, "soft failures" also cause service calls. Soft failures are problems that typically are repaired by the customer, such as a paper jam in a printer. If this type of failure occurs repeatedly, it may eventually cause the customer to call for service, due to frustration about repeatedly having to repair the same problem. Service calls due to hard and soft failures, with the additional observation that some service calls are repeat calls due to the inability of the service person to fix a problem in one visit gives the following model (Equation (5-1)):

unscheduled service calls $= C \times (A \times \text{hard failures} + B \times \text{soft failures})$,   (5-1)

where A through C are coefficients optimized by examining data for similar systems.

Redundancy and Reliability Goals

Both manufacturers and customers would like perfect reliability. However, if reliability goals are set too high, the product may either be delayed or may be too costly. Lower reliability goals, which result in the product being unavailable during failures, seem to be at odds with the requirement that for some assays, it is a medical necessity to have constant access to the assay result. Redundancy is one solution to this problem and plays a role in setting reliability goals. For example, many laboratories have several blood gas analyzers. Apart from providing extra capacity, multiple systems allow for a blood gas result to be produced when one of the systems fails. In most cases, blood gas results cannot be postponed. Redundancy is used even for more highly reliable systems such as electric power. All hospitals have backup generator systems in the case of a power outage.

FRACAS

A definition and use of FRACAS taken from MIL-Handbook 2155 (13) only requires "diagnostic" to be used in place of "military":

"A disciplined and aggressive closed loop Failure Reporting, Analysis, and Corrective Action System (FRACAS) is considered an essential element in the early and sustained achievement of the reliability and maintainability potential inherent in military systems, equipment, and associated software.

The essence of a closed loop FRACAS is that failures and faults of both hardware and software are formally reported, analysis is performed to the extent that the failure cause is understood, and positive corrective actions are identified, implemented, and verified to prevent further recurrence of the failure."

A database containing the following tables facilitates FRACAS deployment for medical diagnostic systems.

- Events. This table contains all failure events, each of which has
  o a failure mode selected from the failure mode table
  o the frequency of the event's occurrence
  o the severity of the event selected from the severity-classification table
  o some usage indicator (calendar time or cycle count)
  o a variety of other demographic data related to the event.
- Failure Modes. This table contains a list of failure modes and an associated corrective action identifying code
- Corrective Actions. This table contains the status and demographic data for each corrective action
- Severity Classification. This table contains weighting factors according to the severity of an event.

Data Analysis

The lion's share of data analysis comes in reviewing FRACAS event records for the latest time period. Each event, or failure, must be assigned to either an existing or a new failure mode. This list of failure modes is developed over time as knowledge grows about how the system fails. The failure mode designation should be as detailed as possible in describing the possible root cause of the failure. Root causes are often different from the symptoms observed. For example, a symptom might be failure to provide a result due to a noise flag, whereas the root cause might be traced to a specific reagent. If the reason for the reagent failure is known (for example, the reagent contains a precipitate), it becomes part of the failure mode. If the root cause is unknown, then "unknown" should be listed in the failure mode's description to separate known causes from unknown causes.

Pareto analysis over a recent time period ranks those failure modes highest that have the highest contribution to the overall failure rate. Each failure mode is quantified for the Pareto by summing the result of the event's severity multiplied by its frequency of occurrence over all events that belong to a failure mode. A typical Pareto analysis in the form of a table and graph (Figure 5-6) is issued regularly to provide management with progress reports.

Corrective Action

The corrective action process uses teams to fix each of the problems at the top of the Pareto chart. Because these problems have the highest impact on the failure

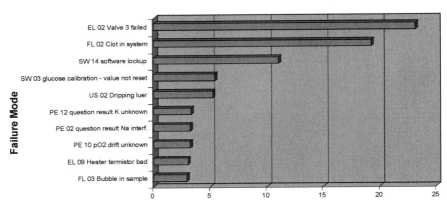

Figure 5-6 Example of a Pareto chart for a blood gas analyzer.

rate and because resources are always limited, it is imperative to marshal resources effectively. Without Pareto analysis, some other selection method might be used to assign people to work on problems such as fixing problems in the order in which they were first reported or perhaps working on a problem that was the most interesting. These selection methods are almost always slower in achieving results in reliability improvement. On a superficial level, an observer would not notice any difference between a traditional reliability program and that of reliability growth management. In each case, the observer would see people working very hard to solve problems.

It was also important to distinguish between a recovery action and a corrective action. Recoveries are actions that repair the system to an operational state as soon as possible, but without root cause analysis and appropriate redesign to ensure that the problem will not recur. For example, a customer who cleared a printer paper jam would have performed a recovery. A corrective action would be to redesign the printer mechanism to eliminate paper jams from occurring.

Measuring Progress

In addition to the Pareto analysis, estimating the speed of fixing problems is a critical success factor in reliability growth management. This can be assessed using Duane plots and the corresponding analysis. In a Duane plot, the log of the cumulative failure rate (y-axis) is plotted against the log of the cumulative time the system is under test (x-axis). This graph often provides a straight line, whose slope is the growth rate (see Figure 5-5). The growth rate is an essential feature of the learning curve phenomenon. Thus, for a system under test, a variety of failures contribute to the overall failure rate. If teams working on solving the top problems fix these problems at a faster rate than new ones occur, a growth rate will be observed. The equation for the Duane relationship is shown in Equation (5-2).

$$\lambda(t) = KT^{-\alpha} \tag{5-2}$$

where λ is the cumulative failure rate at time t
$\quad$ K is a constant
$\quad$ T is the cumulative test time
$\quad$ α is the growth rate

Taking logs of both sides of Equation (5-2) provides Equation (5-3), which is linear and allows one to determine a slope $(-\alpha)$ and intercept (Ln[K]) by linear regression.

$$Ln[\lambda(t)] = Ln[K] - \alpha Ln[T] \tag{5-3}$$

The key parameter estimated from the analysis is the growth rate, $-\alpha$. A large α means a steep Duane curve, which translates to fast reliability improvement. With the two Duane parameters, one can estimate the cumulative failure rate (Equation 5-2) and the instantaneous failure rate (Equation (5-4)).

$$\lambda(t) = (1-\alpha) KT^{-\alpha} \tag{5-4}$$

The cumulative failure rate is an estimate of the average failure rate over the entire time period, whereas the instantaneous failure rate is an estimate of the failure rate at the end of the time period. Equation (5-4) can also be used to predict failure rates for future time periods. This is of key importance to management, which always wants to know when reliability will reach an acceptable level.

There is an important difference between predictions based on reliability growth compared to traditional predictions. To see this, it is helpful to generalize the nature of failures, which can be grouped into three general categories, as follows:

A. Failures that have previously occurred and still occur, for which no fix has yet been found. Some failures may be inherent in components chosen.
B. Failures for which a fix has been proposed, but has not yet been applied to the system.
C. New failures, which might be altogether new or caused by fixes from category B.

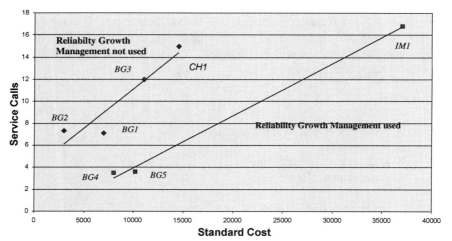

Figure 5-7 Results of reliability growth management. Letters refer to instrument types: BG = blood gas; CH = chemistry analyzer; IM = immunochemistry analyzer.

In traditional predictions, managers commonly count failures in category A, and set to zero failures in categories B and C. That is, all fixes are assumed to be 100 percent effective and without causing any new problems. Moreover, the existence of completely new problems is ignored. One factor that causes this optimism is tremendous pressure by management to speed up the development process. What manager cares to admit that his or her proposed fixes might cause additional problems or might not be completely effective, or that altogether new and unknown problems are lurking in the system? In reliability growth predictions, there is no distinction among categories A through C. The failure-rate predictions are data driven, based solely on the demonstrated rate of fixing problems. Putting things another way, reliability growth predictions are based largely on past data (and the assumption that the past rate of fixing problems will continue), while traditional predictions are based on assumptions about future events.

Results Achieved with Reliability Growth Management

The results of a reliability growth program were compared with previous programs that did not use reliability growth (Figure 5-7). This was facilitated by an assumption that the failure rate of a system is proportional to its complexity. Complexity, in turn, was approximated by standard cost.

The key benefit of using reliability growth management over traditional programs is a shorter development time, achieved by increasing the speed of reliability improvement. This often translates into improved profitability. The reasons for success of the program can be summarized as follows:

- By using a data-driven prediction process, management has an accurate forecast of when the reliability goals will be reached. This allows management to plan for contingencies.
- By developing a metric and a model that counts all failures, coupled with a process to classify and rank problems, the limited available resources can be marshaled most efficiently to work only on problems that most affect the speed of reliability improvement.

REFERENCES

1. Silverberg EC. Predicting project completion. Res Technol Mgmt 1991;34:46–49.
2. Krouwer JS. Beware the percent completion metric. Res Technol Mgmt 1998;41:13–15.
3. Meyer C. How the right measures help teams excel. Harvard Bus Rev 1991;72:94–97.
4. Slater R. Integrated process management. A quality model. New York: McGraw-Hill, 1991.
5. MIL-STD-1629A. Procedures for performing a failure mode effects, and criticality analysis 1980. Washington, DC: Dept. of Defense. Available from *http://www.dodssp.daps.mil/*. Accessed 8 March 2002.
6. Box GEP, Hunter WG, and Hunter JS. Statistics for experimenters. An introduction to design, data analysis, and model building. New York: Wiley, 1978, p. 313.
7. Coleman DE and Montgomery DC. A systematic approach to planning for a designed industrial experiment. Technometrics 1993;35:1–12.
8. Institute of Electrical and Electronics Engineers. IEEE Standard Computer Dictionary: A Compilation of IEEE Standard Computer Glossaries. New York: IEEE, 1990.
9. MIL-Hdbk 189 Reliability Growth Management (1981). US Army Communications Research and Development Command, Fort Monmouth, NJ: 07703.
10. Duane JT. Learning curve approach to reliability monitoring. IEEE Trans Aerospace 1964;2:563–566.
11. KcLain KK. Improving the reliability of diagnostic systems. Medical Device Diagnostic Industry 1994:166.
12. Deming WE. Out of the Crisis. Cambridge, MA: Massachusetts Institute of Technology, Center for Advanced Engineering Study, 1982.
13. MIL-STD-2155 Failure Reporting, Analysis and Corrective Action System (FRACAS). 1995. US Army Communications Research and Development Command Fort Monmouth, NJ 07703.

6

Stage IV: Validation

When you can measure what you are speaking about and express it in numbers you know something about it; when you cannot express it in numbers, your knowledge is of a meagre and unsatisfactory kind. — Lord Kelvin

All models are wrong but some are useful — George Box

WHY MANY PUBLISHED VALIDATION METHODS FALL SHORT IN ASSESSING ASSAY QUALITY

Many published evaluations about assays focus on estimating bias (using linear regression to estimate slope and intercept) and precision (reproducibility). Yet, a survey of *Clinical Chemistry* articles and letters about assay problems showed that complaints rarely mentioned bias and precision as problems, but rather interferences as the causes for clinician complaints (*1*). This chapter shows how traditional evaluation data analysis methods can be modified to better assess assay quality.

VALIDATION METHODS WITHIN COMPANIES

Error Modeling Using Simulation

During product development, huge amounts of data are collected. This data can be used to help construct a model of assay error through simulation. Aronsson et al., provide an example of an error model performed by a hospital lab, although typically these simulations are performed by manufacturers because they have data that is unavailable to labs (*2*). Simulation has also been used to help design labs by modeling work flow (*3*). A benefit of simulation is that it allows one to perform many "what-if" scenarios without the time and cost involved in performing actual experiments. The simulations are particularly useful in setting error requirements. The effort expended in developing an error model is justified for analyzers that will support the development of several assays, although it might also be justified for supporting a single assay to deal with issues after it has been commercialized. The steps required for the construction of an error model follow.

69

A Block Diagram or Flow Chart of the System

To facilitate the construction of an error model, it is helpful to start with a block diagram or a flow chart for all relevant subsystems that will contribute to an assay result. Block diagrams are often prepared separately for instruments and for reagent processes. An abbreviated example of an instrument block diagram is shown in Figure 6-1. Here, the lower set of boxes describe processes that are occurring within the instrument. The boxes above are the actual physical devices that allow the processes to occur. Each physical device has some potential error. For example, a pump must deliver a specified amount of fluid, but the actual amount will vary from sample to sample.

A Cause-and-effect Diagram of Error

A cause-and-effect diagram, also called a fishbone or Ishikawa chart, is often prepared in a brainstorming session. An example of a cause-and-effect diagram for total analytical error, as well as the general requirements for a brainstorming session, was in the previous chapter. A cause-and-effect diagram for an error model is similar to, but more detailed than, the total analytical error chart shown in Chapter 5. The root causes for each error source are added to the cause-and-effect diagram for an error model. For example, a source of imprecision might be caused by the sample pump's inability to deliver exactly the same volume for each sample. An example of a fishbone diagram is shown in Figure 6-2.

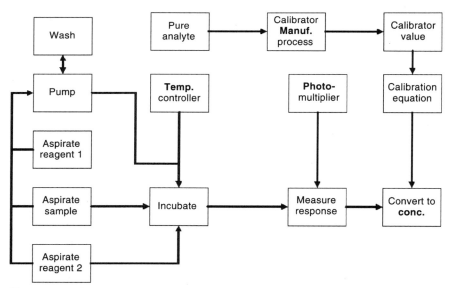

Figure 6-1   Simplified block diagram of an analyzer.

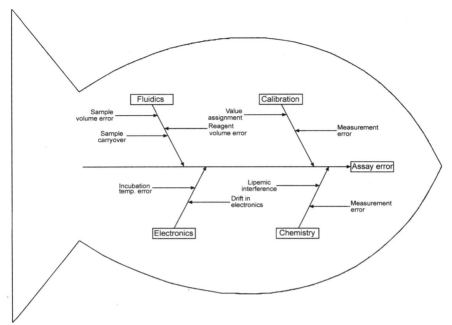

Figure 6-2   Simplified fishbone diagram of assay error.

How the Error Model Works

An error model works by trying to simulate the events in an analyzer, especially those that are subject to error. To do this, a series of inputs are provided to a transfer function, which acts on the inputs to provide the output.

As an example, consider calibration value assignment and its error. The manufacturing group prepares a calibrator and bar codes the calibrator's concentration value they believe they have formulated. The bar code will by read by the analyzer software to provide the calibration equation with the fixed (known) concentration value of the calibrator. Yet, as with most processes, there will be some error associated in attempting to prepare the exact calibrator concentration that matches the bar code value. If the actual concentration value does not match the bar code value, this manufacturing error will propagate a systematic bias according to the calibration equation into every result assayed for that calibrator lot. Of course, the actual error observed for each sample depends on the combined effects of all of the error sources.

If one has collected data on the actual error in manufacturing that has been observed for many calibrator lots, then one can model this error by randomly sampling from this error distribution to simulate the effect of this error source on assay results. The necessary data may need to be collected with experiments on retains. An advantage of simulation is that, in a few minutes, one can simulate running millions of assay results that span many calibrator lots. These data can be analyzed as if they were actually generated in a laboratory. Moreover, with simulated data, one always knows what the true value is supposed to be. This process is repeated for all error sources.

One aspect of simulating error from distributions that is not always appreciated is the way simulation accurately combines probabilities. For example, consider five normally distributed error sources, each with a standard deviation of 2 mg/dL for the error source. For each individual error source, a commonly assumed worst-case 3 standard deviation error would be $\pm$ 6 mg/dL. A worst-case analysis suggests that an error of $\pm$ 30 mg/dL would occur if all individual worst-cases errors occurred at the same time and in the same direction. It would be a mistake to use this information to try to reduce each of the 2 mg/dL error source standard deviations because the likelihood of having all five error sources simultaneously at the same side of three standard deviations is extremely low $(2 \times (0.003)^5 = 4.86^{-11}$ percent). On average, one would need to run 40 billion assays to see one occurrence of this error. Simulation results will reflect these combinations of probabilities.

The simulation results can help with setting manufacturing specifications. The manufacturing process specifies a procedure that both attempts to minimize the error in calibration value assignment *and* to minimize the cost of the process. Simulation allows the manufacturing specifications to be based on:

- the effect of *all* assay errors, taking into account the probability of each error
- the effects of different manufacturing processes such as more testing or different ways of preparing the calibrator, which change the magnitude of an error source

The Simulation Software

Simulation software can be software that is specifically designed to perform simulations, such as Simscript, or a spreadsheet add-in package such as Prism, or a general statistical package such as SAS. The Institute for Operations Research and the Management Science (INFORMS) periodically surveys simulation software (*4*).

Total Analytical Error

Chapter 5 contains a cause-and-effect diagram of laboratory error, which shows how each total analytical error source contributes to total analytical error. It is somewhat surprising that, apart from the above simulation method, estimation of total analytical error is often performed indirectly by attempting to sum up total analytical error components rather than by estimating total analytical error directly. This has changed to some degree with Bland and Altman's publication in Lancet (5), which make their earlier studies (6) available to a wider audience. However, several years after this publication, the direct estimation of total analytical error is not always performed.

The rationale behind the direct estimation of total analytical error is simple. One performs a method comparison experiment and computes the distribution of differences between the new method and the comparison method. Mandel (7) showed how this difference is a combination of random and systematic error (Equation (6-1)).

$$TAE = (y - R) = (y - \mu) + (\mu - R) \tag{6-1}$$

Equation (6-2) is an expansion of Equation (6-1) to account for n replicates of each of m different specimens.

$$TAE = \sum_{j=1}^{m}\sum_{i=1}^{n}(y_{ij} - R_j) = \sum_{j=1}^{m}\sum_{i=1}^{n}(y_{ij} - \overline{\mu}_j) + \sum_{j=1}^{m}(\overline{\mu}_j - R_j) \tag{6-2}$$

where TAE is total analytical error;

y_{ij} is the ith observation from the jth sample of the new method;

R_j is Reference method result for the jth sample; and

$\overline{\mu}_j$ is the mean of the jth sample for the new method.

In Equation (6-2), the second double summation term is a measure of imprecision. The last term represents the distribution of bias that results from each sample's average deviation from reference.

There have been many attempts to combine error sources to arrive at a total analytical error estimate. Perhaps the most common method is to add some multiple of imprecision to bias estimated from a regression equation (8). This simply will not work because Equation (6-2) tells us that we must account for the bias from *each* sample. The bias from a regression equation represents *average* bias across the many possible samples at a specific concentration level. Equation (6-2) does not attempt to average bias across samples. In fact, the last term in Equation (6-2), which represents the random bias across patient samples, has been neglected by most publications that deal with assay evaluations, despite a very complete treatment of this effect (9). The random bias is not simply an academic concern as Krouwer showed using real data for a cholesterol assay. In this case, random bias was the largest total analytical error source (1). More recently, Miller et al., showed how total analytical error is underestimated for an LDL-cholesterol assay by using the National Cholesterol Education Program (NCEP) recommended method of combining imprecision and bias (e.g., bias $+ 1.96 \times CV_T$ (10).

Returning to the direct estimation of total analytical error using Equation (6-2), we first need to distinguish among different classes of assays. There are

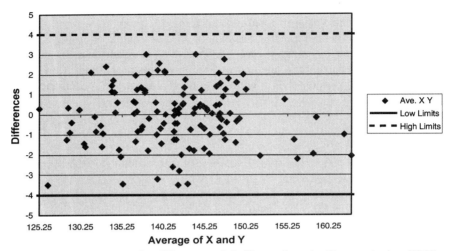

Figure 6-3 Example of a difference plot. (Reproduced with permission EP21-P, "Estimation of total analytical error for clinical laboratory methods; proposed guidlines." NCCLS. 940 West Valley Road, Suite 1400, Wayne, PA 19087, U.S.A.)

definitive, reference, and field methods (11). Definitive methods have the highest accuracy and are performed only by specialized reference labs. Reference methods are also highly accurate, have been validated against definitive methods, and are performed by manufacturers and some hospital labs. Field methods refer to commercial assays.

For many cases in method comparisons, a field method rather than a reference method will be the comparison assay. Here, the bias no longer implies a deviation from truth but rather a difference between the two assays, with no way of knowing which assay (if any) is correct. This issue concerns all methods of estimating bias, not just the direct estimation of total analytical error. One must also be aware that, unless one sufficiently replicates the comparison method (whether it is a field or reference method), the imprecision of the comparison assay will be part of the distribution of differences. This issue applies specifically to the direct estimation of total analytical error. One solution to these concerns, practiced by manufacturers, is to run a three-way (or more) comparison to include the candidate, desired field comparison(s), and reference assays, with sufficient replication in all methods. This study answers the following questions:

1. Which method (comparison or new) is closest to truth (reference)?
2. What will users observe when they switch from the comparison to the new method?

If the answer to question 2 results in differences that are too large, the answer to question 1 may be used to rationalize to users that they are now closer to the true value of the analyte. Ironically, at times manufacturers use these rationalizations to explain differences between assay methods, all of which they themselves manufacture.

There are two common methods of displaying total analytical error (as the distribution of differences) and providing numerical estimates (5, 12).

The plot of differences by concentration (5) shows how biases may differ throughout the concentration range tested. If the comparison method is not a reference method, then the x-axis should be constructed by taking the average between the test and comparison method (6), or else an artificial correlation will result. Note that other plotting variables could be chosen for the x-axis, such as the date the difference was obtained or perhaps the concentration of an analyte other than the one being evaluated. The latter plot would show the effect of an interference. An example of a difference plot is shown in Figure 6-3.

To calculate the total analytical error, one computes the standard deviation of differences multiplied by an appropriate factor corresponding to the desired coverage. For example, multiplying the standard deviation of differences by 2 and adding and subtracting this quantity from the average difference would provide the lower and upper values for a range which would cover about 95 percent of the differences. This is valid only if the distribution is approximately normal. Normality could be achieved by transformation.

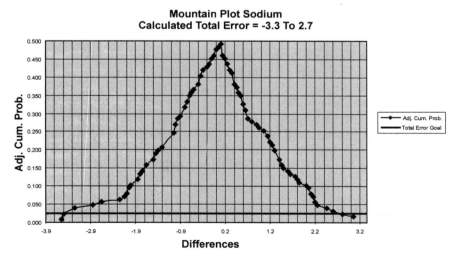

Figure 6-4 Example of a mountain plot. Low and high limits of − 4 and + 4 are not shown on the graph. (Reproduced with permission EP21-P, "Estimation of total analytical error for clinical laboratory methods; proposed guidelines." NCCLS. 940 West Valley Road, Suite 1400, Wayne, PA 19087, U.S.A.)

The mountain plot (*12*) provides an alternate way of displaying the data. It focuses on the tails of the distribution (Figure 6-4). The same percentiles are estimated as in the Bland and Altman method, but the estimation process is non-parametric in that one estimates the corresponding desired percentiles. The advantage of this estimation method is that no distribution assumptions are required.

For either method, one can calculate tolerance intervals (*13, 14*).

Whereas the estimation of total analytical error is a simple and direct way to estimate error from all sources, it does not provide any information about which specific error sources make up the total analytical error that has been calculated. The following methods are designed to assess specific individual error sources.

Multifactor Protocols

Introduction

A common misconception about statistical consultants is that they always want people to collect more data then the amount that was originally planned. Multifactor protocols, which can be used in both industry and hospital laboratories, actually require less data than most people think would be needed to gather the information provided. This saves money and also points out assay problems that one-at-a-time designs sometimes miss.

Multifactor Protocol History

All of the multifactor protocols in this section are based on Cuthbert Daniel's work (*15*). Daniel was a well-known statistical consultant to industry (*16*), who developed multifactor protocols at the request of Dr. Stan Bauer while he was at Technicon Instruments (now Bayer Diagnostics). The original protocols were designed to estimate performance parameters for the SMA and SMAC analyzers and were used extensively to evaluate assays during their development. The protocol for the SMA was also published as an NCCLS guideline (*17*). More recent multifactor protocols were used at Ciba Corning Diagnostics (now also Bayer Diagnostics) and other places, and extend Daniel's designs to take into account features of newer analyzers such as random access capability (*1*).

Understanding Multifactor Protocols

The principles underlying multifactor protocols are simple. If one can understand ordinary linear regression, then one can understand multifactor protocols. In simple linear regression, a model provides a means to estimate the slope and intercept of a set of data (Equation (6-3)).

$$Y = \beta_0 + \beta_1 X + \varepsilon \qquad\qquad (6\text{-}3)$$

where $Y =$ the observation for the method under test
$\qquad \beta_0 =$ the intercept coefficient
$\qquad \beta_1 =$ the slope coefficient
$\qquad X =$ the observation for the reference method
$\qquad \varepsilon =$ error not explained by the model

Another way of looking at this equation is depicted in Table 6-1.

The last term, epsilon, which is often omitted from models but nevertheless plays a role, is equal to the assay imprecision of the test method provided the reference method has no error and the model explains exactly the systematic error in the data. While these might seem to be restrictive assumptions, it is quite possible for a regression model to be a good one (e.g., to nearly explain the data). And if one runs replicates of the reference method, there will be very little error in the average reference result.

In a multifactor protocol, the same basic principles apply. The difference is that one adds more terms than $\beta_0 + \beta_1$. Equation (6-4) describes a multifactor protocol that estimates intercept, slope, linear drift, and random error. In this case, a term, β_2, has been added to account for time of assay since last calibration; therefore, in Equation (6-4), the systematic effect of linear drift is also being estimated. Note that all assays have a time since the last calibration so that all assays could be evaluated by inclusion of this term. This will be discussed later in more detail.

$$Y = \beta_0 + \beta_1 X + \beta_2 t + \varepsilon \qquad\qquad (6\text{-}4)$$

where, in addition to Equation (6-3) $+ \ \beta_2 =$ the linear drift coefficient
$\qquad\qquad\qquad\qquad\qquad\qquad\quad t =$ the time of assay since the last calibration

It is also important to consider the order in which the samples are run. This is a key design factor in multifactor protocols and important even if one does not use a multifactor protocol analysis. For example, a common one-at-a-time sample order for estimation of slope and intercept might be to run the following sequence of samples: L, L, ML, ML, M, M, MH, MH, H, H where L $=$ low, M $=$ mid, and

Table 6-1
Systematic and Random Error Attributes

	Systematic effect	Random effect
Error source is	Known (deterministic)	Unknown (probabilistic)
Y is explained by[a]	$\beta_0 = \beta_1 X$	$+ \varepsilon$
Y is	Known for each sample	Known for the average sample

[a]Assumes that the model is correct.

H = high. The problem with this design is that, if drift does occur, the design will not handle it well. One can show this by calculating the efficiency of the design, as described by Daniel (*15*), which is a measure of how correlated the parameter estimates are. The efficiency is calculated by Equation (6-5).

Efficiency: $\mathbf{P}(i) = 1/(a_{ii} * b_{ii})$ $\hspace{6cm}$ (6-5)

where $\mathbf{P}(i)$ = the efficiency of the ith parameter $\mathbf{P}$
$\hspace{2.2cm}$ a_{ii} = the diagonal term in the matrix $X'X$
$\hspace{2.2cm}$ b_{ii} = the diagonal term in the matrix $[X'X]^{-1}$
$\hspace{2.2cm}$ X = the design matrix

Designs that are uncorrelated have efficiencies of 1.0. The results for the above design are 1, 0.04, 0.04 corresponding to intercept, slope, and drift. This indicates that the terms slope and drift are highly correlated. This compares to efficiencies of 1, 1, 1 again corresponding to intercept, slope, and drift for the design L, ML, M, MH, H, H, MH, M, ML, L. These efficiency calculations can be performed using Excel® software.

$\hspace{0.5cm}$ If linear drift occurs and there is no drift term in the analysis, the first design's slope estimate will be contaminated with drift. This contamination will not occur in the second design, regardless of whether drift is in the model. In addition, the precision of the estimates will be better for the second design. A simulation demonstrated that, with a modest amount of drift, the slope had an estimation error of 8 percent because an efficient experimental design was not used (*18*).

$\hspace{0.5cm}$ So far, we have considered what happens when we run the multifactor protocol once. Using linear drift as an example, how do we know that linear drift will always have the same value in each run? It is more likely to have a *distribution* of values. By running several multifactor protocols, we will be able to assess this distribution.

$\hspace{0.5cm}$ From a modeling standpoint, one could consider the parameter linear drift to be a constant, and that estimation error is the only reason that linear drift values are not the same from run to run. However, it is also possible that linear drift's value—aside from the estimation effect—varies from run to run because the variables that cause linear drift are in turn varying from run to run. A priori, we don't know which of the two cases applies. A wide distribution implies the latter; that is, linear drift is a systematic effect whose value in any specific run can effectively be thought of as random. Thus, in reality, all of the systematic effects discussed below could be classified as random systematic effects. The importance of the "random" qualifier will be known by examining the distribution. Narrow distributions imply little random character. However, if the distribution is wide or especially if the distribution has a wide tail, then randomness is important and is telling us that certain runs may generate outliers. One can also test this formally with analysis of variance (ANOVA) models for these multiple runs.

The residual error (epsilon in Equation (6-4)) in multifactor designs is an estimate of assay imprecision. If this residual error is smaller than the imprecision calculated from a traditional imprecision protocol—testable using an F-test—then there is evidence that the imprecision calculated from these traditional protocols is not purely random error but is inflated with biases such as drift. In a multifactor protocol, the residual error term is closer to purely random error because these biases are part of the model and show up in the bias coefficients instead of inflating the error term.

The software required to provide multifactor protocol parameter estimates requires multiple linear regression, which is widely available and part of most spreadsheet programs.

Table 6-2 shows a selected list of multifactor protocols and the systematic effects that are estimated by the designs.

Readers wishing to construct their own designs should start with Daniel's reference (*15*).

Use and Interpretation of Multifactor Protocols

To properly use multifactor designs, one should examine the technology underlying the assay and, using the cause-and-effect diagram in Chapter 5, should decide which error sources need to be assessed. It is possible that error sources in addition to those in the cause-and-effect diagram play a role. The specific design or series of designs would then be either selected or constructed. One must then plan at what intervals to run the protocols. Finally, one analyzes the runs and presents the results. Whereas it is easy to present textbook cases, let us consider some problem cases.

Examples

Case 1A—The Data Does Not Fit a Simple Model. The following is a constructed case to explain principles and does not use a multifactor protocol. Real cases are not as easy to dissect. In this example, the slope and intercept for an assay are estimated in two separate runs. In run 1, the data fits the model and we get estimates that seem reasonable for slope, intercept, and imprecision. In run 2, the data does not fit the model. But no matter how bad the data is, the equation nevertheless calculates results, although we can add steps to check model fit to warn us of bad fits. Both the slope and the intercept in run 2 are quite different from run 1 estimates, and the error term in run 2 will be higher than that of run 1. Assume that this is because of nonlinearity at the high end of the assay (although at this point we do not know this). The author has seen the following reactions by scientists when something like run 2 occurs:

1. The slope is too low—Let's look at the data (often by looking at a plot).
2. The slope is too low—We need to fix the system by doing *a, b,* or *c.*

Table 6-2
Selected Multifactor Designs

Ref.	Sample details		Bias		Carryover		Non-linearity	Drift		Drift × Level	Interferences
	Levels	Length	Prop.	Const.	Sample	Reag.		Lin.	Quad.		
13	3	9	x	x	x		x	x			
13	5	46	x	x	x		x	x	x		
16	3	11	x	x		x	x	x			
16	3	27	x	x	x	x	x	x	x	x	
36	3	9	x	x	x		x	x			x

3. These protocols don't work—Stop running them.
4. Stop running high samples.

Solution 1 is an appropriate response. The real problem with solution 2 is that the results have been accepted *without* the proper scrutiny. One must always remember that the parameter estimates are a solution to a model, and the choice of model may not be appropriate. Solution 3 reflects the resistance that occurs sometimes when scientists are asked to deviate from their regular procedures, although this case involves only simple regression. Someone will always suggest Solution 4.

Case 1B—The Data Does Not Fit a More Complicated Model.

Although Case 1A was constructed, real cases occur in which the model is incorrect yet still provides useful information. In one case, a tobramycin assay exhibited consistent negative sample carryover, using a multifactor protocol similar to EP10 (*17*). Because of the analyzer design, sample carryover was highly unlikely. Thus, one could say that the model was wrong. However, it is important to understand what happens in an incorrect model. As mentioned previously, the software will still estimate all parameters, and it does so mathematically, with no knowledge of what is happening physically. In this case, a systematic effect was mathematically similar to the physical phenomenon of sample carryover. The model calculated some of this error in the slot called sample carryover and put the rest of the error in the imprecision error term. Scientists were able to propose and verify a hypothesis—namely that the drug tobramycin was sticking to the sample probe and subsequently washing out. This could be shown to resemble the effect of sample carryover mathematically, albeit with a negative sign. The message is not that one should neither accept results that don't make sense nor discard the results of the analysis, but that one should instead consider results like these a red flag, requiring further study. Something was occurring that was detected by the multifactor protocol and needed to be explained. The multifactor protocol could thus be considered a screening protocol. Although it was based on an incorrect physical model, it was nevertheless good at trapping error sources because of other physical causes.

A similar case was discovered for an ion selective electrode (ISE)-based lactate assay that exhibited negative carryover. In this case, hysteresis was proposed as an explanation of the error source.

The screening idea can be applied to large numbers of runs by using software that combines all runs into tables that flag specific runs that deviate from limits.

Case 2—Use of a Multifactor Protocol in Manufacturing.

A random access analyzer was being manufactured. In the analyzer design, the same probe was used to dispense either reagent or patient sample. In a random access analyzer, any assay can follow any other assay. This requires that there be no carryover from reagents used in a previous assay to the current assay. One such troublesome sequence of assays is an aspartate aminotransferase (AST) assay followed by lactate dehydrogenase (LDH). This is because LDH in high concentration, relative to that of a patient sample, is used in the formulation of AST. Carryover of AST reagent into an LDH assay could yield falsely high results. To ensure that each instrument was manufactured with an acceptable negligible amount of reagent carryover, a 12-sample-long multifactor design was constructed and run as part of the release criteria for each instrument (*19*).

Implementation

A common source of resistance to using multifactor designs by development staff is that interpretation of the results is not as simple as with one-at-a-time designs. One cannot simply glance at the data and tell what is going on. Thus, some scientists are more comfortable with one-at-a-time analyses. The resistance will be expressed in many possible ways including the following:

- The design and analysis are too complicated
- I understand it, but the models won't work for our data

To help overcome resistance, one must point out to management the cost savings possible by reducing the data required by multifactor protocols. In one comparison, it was shown (*18*) that one-at-a-time protocols required 5 times more samples to be run than an equivalent multifactor protocol to provide the same information. The one-at-a-time protocols actually require more than 5 times more work because each one-at-a-time protocol must be set up and analyzed, requiring technician and analyst time.

An additional solution to resistance is to provide a training session so that scientists will feel more comfortable in analyzing the results. Because scientists are often ultimately responsible for the assay's design quality, it is understandable that they object to using multifactor protocols if they are not comfortable with them.

Additional Special Studies

There are additional error sources for some assays that require separate estimation to ensure quality for all aspects of assay use.

Diagnostic Accuracy

In Chapter 5, the box above total analytical error is an assay's most important attribute: diagnostic accuracy. A good assay has high sensitivity—the proportion of diseased patients that test positive and high specificity—the proportion of non diseased patients that test negative. For many assays, the diagnostic accuracy has already been established. Review and standards for establishing diagnostic accuracy continue to evolve. (*20–22*).

The Detection Limit

The detection limit is traditionally defined as the lowest concentration at which analyte can reliably be determined as present. For some assays, the detection limit (minimum detectable dose) is the most important performance parameter. This is often the case for drugs of abuse, tumor markers, and infectious disease assays. Here, assays errors that cause false positive or false negative results can have dire consequences.

Specific Interference Studies

In a method-comparison experiment, interferences may be estimated indirectly as a random error component. However, a particular interfering substance may or may not be present by chance in the samples being evaluated. Moreover, the random interference metric provides no information about the nature of the interfering substances. One can directly study specific potentially interfering substances, which will conceptually characterize the individual error for all interfering substances, *provided* one has identified all potentially interfering substances. Judging from complaints from publications, there appear to be cases in which the effect of an interfering substance has not been correctly identified by a manufacturer. In a recent example, a serious interference was found for a digoxin assay, caused by a drug that was not approved in the United States (*23*). The interference resulted in a falsely low reported value of digoxin, which resulted in the patient receiving an overdose of digoxin.

Direct vs. Indirect Methods of Estimation

The above example about estimating interferences illustrates that there are often different ways to estimate a parameter. Each has its advantages and its limitations. As another example, nonlinearity can be estimated by a lack of fit test for a linear regression equation that estimates slope and intercept (*24*) or by a series of lack of fit tests that compare fits to polynomial equations of different powers with all points included or some points excluded (*25*). The first approach provides a global detection for nonlinearity, whereas the second method, which is more complicated, provides the location and magnitude for any nonlinearity that has been detected.

Estimation of Outliers

In the study mentioned at the beginning of this chapter and based on a year of *Clinical Chemistry* articles, most assay performance complaints were about interferences (*1*). In a typical case (*26*), a clinician complained to the lab about a reported result because it did not make sense when viewed along with the other information available to the clinician. The laboratory reran the sample with an assay from a different manufacturer and confirmed that the originally reported result was incorrect. The clinical chemists then developed and tested a hypothesis to explain the erroneous result and, in cases in which the hypothesis proved correct, the laboratory published the result in *Clinical Chemistry*. In some outlier examples, the reported result was known to be incorrect, but the mechanism of interference remained unresolved and thus remained unpublished.

Outlier results are the most dangerous performance problems because they are a likely laboratory cause for a clinician to make an incorrect treatment decision. An example of this was described in Chapter 1, where repeated hCG outliers resulted in unnecessary surgery.

This makes it even harder to explain why estimation of outlier rates or even mention of them is rare in evaluation studies. Perhaps this is because outlier studies are difficult to perform. It is not easy to define outlier goals or to prove that outliers have a low rate of occurrence. This section addresses these issues.

Outliers can be viewed as belonging to several types: mistakes, interferences, instrument and reagent problems, and statistical outliers.

Mistakes — A mistake such as labeling a sample with the wrong patient name can cause an outlier that is unrelated to the analytical process (*27, 28*). This is often called a preanalytical error. Postanalytical errors are also possible.

Interferences — Although manufacturers study many candidate-interfering substances, there is no guarantee that they have correctly identified all substances that actually interfere. Laboratories may miss the warnings about interfering substances that the manufacturer puts into the product literature.

Instrument and reagent problems — Each assay result arises from the execution of thousands of events that occur within the instrument reagent system. If one of these processes behaves abnormally, an outlier may result. This outlier could occur during the assay of a single patient sample, which would most likely escape routine quality control procedures. Manufacturers try to prevent the occurrence of such outliers by embedding signal processing algorithms in the result calculation. When a problem signal is detected by these algorithms, the result is not reported.

Statistical Outliers — Although extremely rare, it is possible that the system is behaving normally and that an outlier result is simply caused by statistical variation.

Outliers can also be classified by the magnitude of their deviation from truth. As a value deviates more and more from truth, at some point the magnitude of the deviation is great enough to become a medically important outlier. As the deviation increases even more, the deviation will be so large as to be an unbelievable result and in some cases will even be medically or physically impossible. These results, while still technically outliers, are unlikely to cause harm.

Outlier Goals

Much has been written about medically acceptable results, with one example being the Clarke diagram for glucose (*29*). The rationale for setting limits that define medically acceptable results is that these limits are intended to protect clinicians from making incorrect treatment decisions when errors in a lab result become "large." The thinking is that, if all the results fall within the medically allowable limits, few or no incorrect treatment decisions will result from lab error. If results fall outside limits, then incorrect treatment decisions are likely to occur. Of course, clinicians will react differently as each clinician forms a clinical picture of the patient and then receives the incorrect assay result.

These medically acceptable limits are usually meant to be associated with total analytical error, not outliers. Total analytical error covers a high percentage of result differences—usually 95 percent—from truth. If an assay were just to meet its total analytical goal, this would mean that 5 percent of results would be outside total analytical error. A busy lab commonly reports 1 million or more results per year, which implies that in the worst case, 50,000 results per year are beyond the total error goal for an acceptable assay. Even if a 99 percent total analytical error goal were used, 10,000 unacceptable results would still be observed. Another set of limits to guarantee that results that are outside total analytical error are still reasonable is the concept behind outlier limits. Once we have a set of limits for outliers, we still must set a rate for the percentage of observations that can be outside these limits for an assay. Knowing that a single result outside these limits can cause death or injury, one may be tempted to specify an outlier rate of zero. In fact, the president of a medical diagnostic corporation, knowing that certain software bugs cause outliers, once told his audience that the target rate for software bugs for a new system was zero. The problems with a zero rate were covered in Chapter 4, with the main issue being that a zero rate is impossible to verify; one cannot prove the null hypothesis.

Specifying an outlier rate greater than zero does not mean that an outlier rate greater than zero will be observed. It is possible that no outliers will ever be observed, but now one will be able to evaluate assays for this important parameter.

Estimation of Outlier Rates

The Problem With Statistics Based on the Normal Distribution.
Consider the following data set: {0, 0, 0, 0, 0, 0, 0, 0, 0, 1,000,000}. If one uses the rule to reject a bad value—reject the suspected value if its Z-score is 3 or greater (calculated by $(x_i - \bar{x})/s$)—the value of 1 million will not be flagged as an outlier in this data set of 10 observations. This is because it is impossible to obtain a Z-score of 3 (or greater) for 10 observations, regardless of what the 10 numbers are (30). One could correctly identify the suspicious value as an outlier by comparing its Z-score (2.846) with a table of critical values developed by Grubbs (31). Perhaps more important is the assumption that the data are normally distributed. This assumption is difficult to prove, and if it is not true, Z-scores will not be appropriate anyway. Moreover, we are not really interested in

Table 6-3
Combined Total Analytical Error and Outlier Assessment

Case	Total analytical error	Outlier rate	Outcome
1	Acceptable	Acceptable	Pass
2	Acceptable	Unacceptable	Fail
3	Unacceptable	Acceptable	Fail
4	Unacceptable	Unacceptable	Fail

a statistical definition of an outlier. We simply want to count the frequency of results outside preset outlier limits with those limits based on medical, not statistical, reasons. The combined assessment of total analytical error and outlier possibilities is shown in Table 6-3.

Case 2 is a situation in which the required percentage (95% or 99%) of the error distribution fits inside the total analytical error limits but there are too many points outside outlier limits. In Case 3, the required percentage (95% or 99%) of the error distribution does not fit inside the total analytical error limits, yet the number of points outside outlier limits is below the number required to fail the outlier rate specification.

Either Case 2 or Case 3 will cause problems. Case 2 can be corrected by identifying and preventing outliers. Case 3 requires improving assay imprecision and in quality terms is an assay that is not capable (see Chapter 7). Case 3 also presents a challenge for a quality-control program because one must set controls limits to be wider than medically acceptable error to prevent too many alerts. If one were to use more narrow quality-control limits, one could do nothing in the case of an alert.

Outlier Estimation Based on the Binomial Distribution. One can always base outlier rate estimates on discrete event statistics. Here, each result can be considered either to be an outlier, or not. The number of results that fall outside the outlier limits, divided by the total number of samples run, is the outlier rate.

Table 6-4 shows the number of observations required to show the maximum percentage outlier rate with the stated level of confidence (*32*). To correctly report the results of such a study, one must say something such as "Outliers, if present, will be at or below the stated rate." That is, one has not proven that there are no outliers but only that if they do exist, they will be at or below the stated rate.

While these large sample sizes might seem to be out of reach of most hospital labs, manufacturers do perform studies of these sizes during the product development cycle. Hospital labs can use quality-control values to estimate these rates, although in this case, the estimated rate will probably underestimate the true rate because patient samples are not included.

Table 6-4
Maximum Percentage Outlier Rates for Different Sample Sizes

Sample size	Number outliers found	Maximum percent outlier rate (95% confidence)	Maximum percent outlier rate (99% confidence)	ppm outlier rate (95%)	ppm outlier rate (99%)
10	0	25.9	36.9	259,000	369,000
100	0	3.0	4.5	30,000	45,000
1,000	0	0.3	0.5	3,000	5,000
1,000	1	0.5	0.7	5,000	7,000
10,000	0	0.03	0.05	300	500
10,000	1	0.05	0.07	500	700
10,000	10	0.2	0.2	2,000	2,000

Detection of Outliers. One reason outliers do not cause as many medical problems as one might expect is that laboratories have a means to detect outliers before they are reported to the clinician. If outlier results are reported, clinicians will question some results before acting on them. Typically, assay results are validated before they are released to a clinician, and this validation process has detection strategies that are explained in this section. Expert systems hold the most promise in detecting outliers; however, their optimal use relies on a complete electronic patient record. For example, in the case of a drug that has been prescribed for a patient, an expert system that uses an electronic patient record could do the following:

- Determine whether the drug has actually been given and its dose;
- Check the drug against the list of interfering substances for the assay;
- Adjust a delta check algorithm to account for the effect of the drug on the analyte that is being assayed

Systems that have been useful in detecting outliers follow:

Reflex Testing—Many labs set up criteria for automatically retesting a result that is unusual, such as one that is abnormally high or abnormally low.

Delta Check—This algorithm compares serial results for the same patient sample (*33*). Using historical data, one can compute the probability of the size of differences.

Multivariate Analysis—Here, algorithms analyze the likelihood of combinations of two or more assays results (*34*).

Expert Systems—These comprise a broad class of software systems that provide algorithms that use several outlier detection criteria (*35*), such as the drug example above. Rule-based systems formalize in software the process used by laboratorians to validate results.

Clinician Detection—When a clinician receives a lab result, he or she often has some expectation of the value's expected range. This range can be considered a surrogate reference method. For example, consider a hyperglycemic patient with a glucose of 350 mg/dL who has just been given an appropriate amount of insulin. The next glucose assay result is expected to be lower, around 80–130 mg/dL. If this is not the case, then the clinician will try to determine the reason, including the hypothesis that the lab result is in error.

EXTERNAL (CUSTOMER) VALIDATION METHODS

Why Manufacturer Trials Held at Customer Sites Often Fail to Detect Problems

Virtually all products are field tested before they are launched. Oftentimes management has high expectations that these external trials will demonstrate both a new product's improvements as well as point out any remaining problems. Management's rationale is "Let's get the product in the hands of the customer." But ironically, these trials often fail to provide much new information. The reason is that despite a new assay being in "the hands of the customer," most hospital labs do not run an unreleased assay in the same way that they would run an assay in routine use; hence, they don't act as real customers. There are several reasons for this. Most hospital labs have optimized their physical and human resources. There is often neither the space nor the technicians available to participate in evaluations of new assays. Labs do have an incentive to participate, because they are well compensated for performing evaluations. However, they often perform the evaluations in areas apart from the main lab, and by people who do not routinely run the assay currently used to report patient results. The fact that the assay under evaluation is not being used to report patient results means that many lab procedures are bypassed. For example:

- Because assay results are not subjected to the lab's criteria for releasing results to clinicians, no one looks at the results (patient or quality control);
- There is no clinician-based time pressure to produce assay results;
- Because the results are not reported to clinicians, there is no opportunity for complaints on result quality.

Of course, manufacturers try to ameliorate this situation by ensuring—by writing into the protocol—that the assay is run by the usual lab technicians within their normal work flow and that these personnel are given criteria to review the results.

REFERENCES

1. Krouwer JS. Estimating total analytical error and its sources. Arch Pathol Lab Med 1992;116:726–731.
2. Aronsson T, de Verdier CH, and Groth T. Factors influencing the quality of analytical methods—A systems analysis with use of computer simulation. Clin Chem 1974;20:734–738.
3. Vogt W, Braun SL, Hanssmann F, et. al. Realistic modeling of clinical laboratory operation by computer simulation. Clin Chem 1994;40:922–928.
4. *http://www.lionhrtpub.com/orms/surveys/Simulation/Simulation.html*. Accessed 17 February 2002.
5. Bland JM and Altman DG. Statistical methods for assessing agreement between two methods of clinical measurement. Lancet 1986;327:307–310.
6. Altman DG and Bland JM. Measurement studies in medicine: The analysis of method comparison studies. The Statistician 1983;32:307–317.
7. Mandel J. The statistical analysis of experimental data. New York: Dover, 1964, pp. 104–105.
8. Petersen PH, Stöckl D, Westgard JO, Sandberg S, Linnet K, and Thienpont L. Models for combining random and systematic errors. Assumptions and consequences for different models. Clin Chem Lab Med 2001;39:589–595.
9. Lawton WH, Sylvester EA, and Young-Ferraro BJ. Statistical comparison of multiple analytic procedures: application to clinical chemistry. Technometrics 1979;21:397–409.
10. Miller WG, Waymack PP, Anderson FP, Ethridge SF, and Jayne EC. Performance of four homogeneous direct methods for LDL-cholesterol. Clin Chem 2002;48:489–498.
11. Tietz NW. A model for a comprehensive measurement system in clinical chemistry. Clin Chem 1979;25:833–835.
12. Krouwer JS and Monti KL. A simple graphical method to evaluate laboratory assays. Eur J Clin Chem Biochem 1995;33:525–527.
13. Estimation of total analytical error for clinical laboratory methods. Proposed Guideline NCCLS EP21P. NCCLS, 771 E. Lancaster Ave., Villanova, PA, 2002.
14. Hahn GJ and Meeker WQ. Statistical intervals. A guide for practitioners. Wiley: New York, 1991, pp. 58, 90.
15. Daniel C. Calibration designs for machines with carryover and drift. J Qual Technol 1975;7:103–108.
16. *http://www.amstat.org/publications/tas/abstracts_98/hunter.html* Accessed 17 February 2002.
17. Preliminary evaluation of quantitative clinical chemistry laboratory methods. Approved Guideline NCCLS EP10A. NCCLS, 771 E. Lancaster Ave., Villanova, PA, 1998.
18. Goldschmidt HMJ and Krouwer JS. "Multifactor Experimental Designs for Evaluations," in Evaluation Methods in Laboratory Medicine. R Haeckel, ed., VCH, Weinheim, Germany, 1993.
19. Krouwer JS, Stewart WN, and Schlain B. A multifactor experimental design for evaluating random access analyzers. Clin Chem 1988;33:1894–1896.
20. Reid MC, Lachs MS, and Feinstein AR. Use of methodological standards in diagnostic test research. Getting better but still not good. JAMA 1995;274:645–651.

21. Assessment of the clinical accuracy of laboratory tests using receiver operating characteristic (ROC) plots. Approved Guideline NCCLS GP10A. NCCLS, 771 E. Lancaster Ave., Villanova, PA, 1995.

22. Standards for Reporting of Diagnostic Accuracy (STARD) *http://www.consortstatement.org/stardstatement.htm.* Accessed 18 March 2002.

23. Steimer W, Müller C, and Eber B. Digoxin assays: Frequent, substantial, and potentially dangerous interference by sprionolactone, canrenone, and other steroids. Clin Chem 2002;48:506–516.

24. Evaluation of the linearity of quantitative analytical methods. Approved Guideline NCCLS EP6P. NCCLS, 771 E. Lancaster Ave., Villanova, PA, 1986.

25. Evaluation of the linearity of quantitative analytical methods: A statistical approach. Approved Guideline NCCLS EP6P2. NCCLS, 771 E. Lancaster Ave., Villanova, PA, 2001.

26. Cartier LC, Leclerc P, Pouliot M, Nadeau L, Turcotte G, and Fruteau-de-Laclos B. Toxic levels of acetaminophen produce a major positive interference on Glucometer Elite and Accu-chek Advantage glucose meters. Clin Chem 1998;44:893–894.

27. Witte DL, VanNess SA, Angstadt DS, et al. Errors mistakes, blunders, outliers, or unacceptable results: How many? Clin Chem 1997;43:1352–1356.

28. Hinckley CM. Defining the best quality-control systems by design and inspection. Clin Chem 1997;43:873–879.

29. Clark WL, Cox D, and Gonder-Frederick LA, et al. Evaluating clinical accuracy of systems for self-monitoring of blood glucose. Diabetes Care 1987;10:622–628.

30. Schiffler RE. Maximum Z scores and outliers. Am Statistician 1988;42:79–80.

31. Grubbs FE and Beck G. Extension of sample sizes and percentage points for significance tests of outlying observations. Technometrics 1972;14:847–854.

32. Hahn GJ and Meeker WQ. Statistical intervals. A guide for practitioners. Wiley: New York, 1991, p. 104.

33. Wheeler LA and Sheiner LB. A clinical evaluation of various delta check methods. Clin Chem 1981;27:5–9.

34. Oosterhuis WP, Ulenkate HJLM, and Goldschmidt HMJ. Evaluation of LabRespond, a new automated validation system for clinical laboratory test results. Clin Chem 2000;46:1811–1817.

35. Fuentes-Arderiu X, Castiñeiras-Lacambra MJ, and Panadero-Garcìa MT. "Evaluation of the VALAB expert system." In Eur J Clin Chem Clin Biochem 1997;35(9):711–714.

36. Krouwer J and Monti K. Modification of NCCLS EP10 to include interference screening. Clin Chem 1995;41:325–326.

7

Stage V: Commercialization

Ship it! — Music to a commercial manager's ears

HOW CLAIMS DIFFER FROM INTERNAL SPECIFICATIONS

During product development, the set of performance specifications that scientists and engineers are working to meet are generally not available outside the company. Even after they have been met, these "internal specifications" are kept confidential for a variety of reasons. They may be considered proprietary, or they are presented in a way that may be unappealing from a marketing perspective, such as being stated too technically. At some point, a publicly available set of performance specifications, which now become claims, are issued. This task is often performed by the marketing group (*1*). This set of claims is often markedly different from internal performance specifications. Two common formats for publicly issued claims follow.

The Typical Data Claim

The typical data claim is usually based on clinical trials and bases the performance claims on the average results from these trials. Although many customers will experience performance "close" to the claim, half the customers will have better performance, and half worse performance, than the claim. To determine whether a customer's performance is within the claim, a statistical test must be performed. A "typical data" performance claim will always appear more favorable to a marketing department than a "guaranteed" performance claim. Conversely, the service department may receive complaints from customers who experience performance worse than the stated claim, even though there may be nothing wrong with these systems. Asking a customer to statistically test their observed performance adds complexity to an evaluation. Companies realize this situation, and many offer a service to evaluate an installed system to determine whether it is performing within claims.

The Guaranteed Performance Claim

The guaranteed performance claim provides a limit that everyone is expected to obtain, or else the service department should be contacted. This simplifies customer evaluations because no statistical test is required. One can never be sure what the basis is for the guaranteed performance claim. In fact, it is possible to arbitrarily provide a guaranteed performance claim or to provide claims that more than a few customers will fail. A statistically based claim might be based on a tolerance interval, which provides a specified percentage of confidence that a specified percentage of a population will be contained with stated limits.

The two different types of claims are illustrated in Figure 7-1.

This figure is similar to Figure 4-3 from Chapter 4. In Figure 7-1, both Results A and B meet the typical and maximum error limit claims, although Result B may cause a customer complaint because it has a bias and must be statistically tested to determine whether the bias is within the typical data specification. Result C fails both the maximum error limit claim and the typical data claim, but again a statistical test must be performed to prove that result C fails the typical data claim.

An NCCLS guideline recommends that, while the presentation of both types of claims is helpful, the maximum error limit claim must be presented, with the typical data claim being optional (2). The rationale is that it is more important to know whether an assay meets minimum performance rather than to know that its performance is "typical."

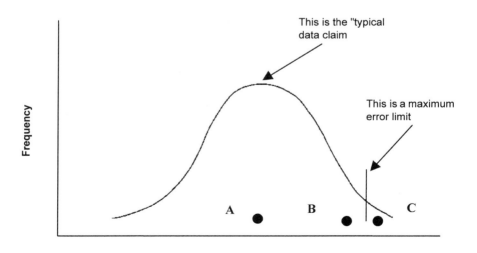

Figure 7-1 Typical and guaranteed performance claims.

OBTAINING AND ANALYZING REMOTE DATA

Several years ago, the following situation would be an example of future technology to improve service rather than a description of an event that has actually occurred.

An Airbus 340 with 4 General Electric (GE) engines is 1 hour into its 11-hour flight from Hong Kong to New Zealand. Inside one of the engines, small bits of insulating skin peel off and fly out the back. The breached surface lets in cold air in, which causes the temperature to drop. The pilots are unaware of this situation. Three hours of temperature data recorded by thermocouples within the engine compartment are uploaded to a satellite, which relays the information to a computer at a GE site near Cincinnati. This computer analyzes the temperature data, previous failure patterns, and the airplane's maintenance records to correctly identify the problem as skin delamination in the engine's thrust reverser. The airline is notified, and when the plane lands, maintenance workers repair the problem without any delay to the schedule (*3*). Had the delamination been allowed to continue until it was noticed by visual inspection, the plane would need to have been taken out of service for a lengthy repair.

Nowadays, this technology is a reality and is also being used by the medical diagnostics industry. Achieving the level of competence in the GE example requires development of an instrument system diagnostic strategy. The strategy, which will affect early as well as later stages in the product development cycle, entails the following:

- fault detection
- data collection
- data analysis

Fault Detection

This involves the use of fault trees and FMECAs to define failure modes and fault detection methods. BIT (built-in test) is a common detection method that often requires a sensor and associated software to track a key parameter. All data collected by these sensors is usually stored in a database within the instrument system. This data is of little value to the customer, who would not understand it anyway, but it is also of little value to the manufacturer unless the manufacturer has easy access to it.

Data Collection

Figure 7-2 shows a typical architecture that is used to collect data from individual instruments to a centralized data store. As technology evolves, the internet will replace the modem.

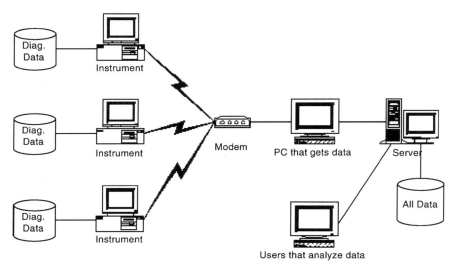

Figure 7-2 Typical data-collection architecture.

Data Analysis

As in the GE example, the ideal case is to alert service of potential problems before they affect customers. Neural networks (4) and classification and regression trees (5) are data-analysis methods used to analyze instrument diagnostic data sources.

Not only can one use data obtained from sensors buried deep within the instrument, one can also use the quality-control data that is regularly collected by the customer for each instrument. The advantage of keeping a centralized quality control data store is that the manufacturer has additional information unavailable to the user. That is, performance of the specific reagent lot from manufacturing records and the comparison of a specific instrument's performance to the population of instruments can be performed by a manufacturer, but not by an individual customer.

The use of quality-control data versus the data from a variety of diagnostic sensors is similar to the discussion in Chapter 6 of direct versus indirect methods of assay error estimation. Quality control in this case is an indirect method; it raises an alert that something is wrong but provides little information about the source of the problem. The diagnostic sensor data is more direct in pinpointing the actual cause of the problem.

The previous technique used realtime sensor or quality control data to prevent failures. The following technique uses failure data for process improvement.

The Mean Cumulative Repair Function

Nelson provided a graphical method that can be used on repairable systems data (6). In this type of graph, the y-axis displays the mean cumulative number of repairs, although one could use the mean cumulative cost of repairs instead. The *x*-axis displays a usage function, which for instrument systems is often the number of cycles. For some systems, cycles are so closely related to time that calendar time (for example, days in service) can be used. A constructed example of this graph is shown in Figure 7-3.

In this example, the repair rate for this instrument system markedly decreases after 25 days. This indicates that, if the repair rate was not isolated to an issue that could be easily remedied, it would make sense to burn in the systems in-house for about 25 days rather than to have customers experience a high failure rate. In general, this analysis provides insight to accomplish the following:

- Determine whether the repair rate increases or decreases with system age (useful for system burn-in or retirement decisions);
- Reveal unexpected information from the plot;
- Compare different designs;
- Predict future repair costs.

The reliability growth methods discussed in Chapter 5 can also be used to monitor reliability progress for systems that have been released. What Duane plots often show after a number of years is that the reliability growth has ended and the curve is flat. Of course, growth will stop if teams are not working on problems. However, another reason for growth to stop is that inherent reliability

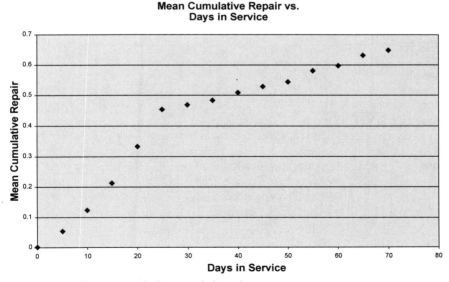

Figure 7-3   Mean cumulative repair function.

of the system has been reached. Training has ensured that both customer use and service personnel no longer cause problems. There are no remaining issues that can be fixed by a practical design change. Figure 5-6 in Chapter 5 presented the results of reliability growth for newly released systems. The corresponding figure for mature systems would have showed much better reliability.

The following technique uses quality control data and can be performed either by a manufacturer or by an individual customer.

USING PROCESS CAPABILITY TO COMPARE PERFORMANCE ACROSS ASSAYS

Although quality control has been a long-standing and useful practice in diagnostic laboratories, few attempts have been made to try to compare the quality of different assays. Process capability statistics represent one means of doing this because they produce a unitless measure that can be used across assays. The process capability metric C_{pm}, also called the Taguchi Capability Index, is ideal for this purpose because it combines systematic and random errors much like the total error concept with which laboratory workers are familiar. This section explains how this metric can be used with existing quality-control data to compare quality across assays.

Process capability statistics have been in existence for some time (7–9) and include C_p, C_{pk}, C_{pm}, and others. C_p is a measure of variability, whereas C_{pk} and C_{pm} measure bias from target in addition to variability. Because C_{pm} has features that make it easier to use than C_{pk}, all discussion will use C_{pm}. The formula for calculating C_{pm} is

$$C_{pm} = \frac{(\text{USL} - \text{LSL})}{6\tau} \tag{7-1}$$

where

$$\tau = \sqrt{(1/n)\sum(x_i - T)^2} = \sqrt{\sigma^2 + (\bar{x} - T)^2} \tag{7-2}$$

and

$$\begin{aligned}
\text{USL} &= \text{upper specification limit} \\
\text{LSL} &= \text{lower specification limit} \\
\tau &= \text{total analytical error} \\
\sigma &= \text{standard deviation} \\
T &= \text{target}
\end{aligned}$$

Several assumptions are traditionally required for process capability statistics. First, the process (e.g., the assay) must be "in control." Also, the data must be either normally distributed or from some known distribution. If these assumptions have been met, capability indices in general provide information about how many results will fall outside specification limits. C_{pm} provides information about how close assay results are to target.

For actual laboratory data, it would be difficult to guarantee that each assay is always in control, assuming one could decide what "in control" means. Additionally, quality-control results may not be distributed normally. The rest of the section will show how this metric can be used without satisfying either that the process is in control or that the data is normally distributed.

The Difference Between Quality Control and Process Capability

Traditional quality control is used to detect changes in the process—that is, the process that yields assay results. Process changes, when detected, act as a warning system to alert users that unless action is taken, the results from the suspect run as well as assay results in future runs may deviate in unpredictable ways from historical data. However, results from out-of-control assays, while certainly not desirable, might not result in errors in medical decisions. That is, results that fail quality-control limits may still fall within certain other limits, referred to here as medically acceptable error limits, provided of course that the medically acceptable error limits are wider than the quality-control limits. It would be a mistake to simply widen quality-control limits to the medically acceptable error limits when performing quality control. While this would produce fewer alerts, the likelihood of out-of-control processes would be higher, leading to a higher likelihood of unpredictable results and potential medical errors. Unfortunately, while all labs have quality-control limits, not all labs have medically acceptable error limits. For an example of medically acceptable error limits, see the list developed by Brigden and Heathcote (*10*). There is of course no guarantee that medically acceptable limits will be wider than the laboratory's quality-control limits.

The strategy presented here recommends leaving quality control as-is and measuring process capability as an additional metric. Process capability provides an estimate of total analytical error that is scaled by medically acceptable error limits. Because one is using quality-control materials, not all elements of total error are included in this estimate. Lawton and co-workers (*11*) showed that random interferences are an important error source and can only be estimated by using patient samples.

An Example Set of Assays Compared

The input data for this method are all quality-control data collected for each assay for a specific time interval. One must decide on data exclusion rules. Possible choices are the following:

- Exclude only those out-of-control results that have caused all patient results to be rerun for the associated time span of the out-of-control period;
- Exclude mistakes, such as running the wrong quality control concentration level;
- Include all results (exclude nothing).

One can of course calculate the metric for different exclusion rules. Because the metric is intended to provide information about the current state of assays, only recent data should be used. We recommend a sliding window that includes the last three months of data. In the United States, this provides a minimum of 90 observations per assay per level of quality control. An example process capability ranking for selected analytes is shown in Table 7-1, with two associated mountain plots shown in Figure 7-4. These data were supplied by Dr. Henk Goldschmidt, although in this example, data and limits were changed for purposes of illustration. Besides C_{pm}, using Equation (7-2), we report the percentage of total error due to bias and imprecision, and the percentage of observations, if any, that fall outside medically acceptable limits. We also prepare a mountain plot for each C_{pm} that is calculated. The table is presented with C_{pm} data sorted in ascending order. Assays with the lowest C_{pm} values or assays that have observations outside specifications suggest further examination.

Interpretation

Remembering that we are violating the standard assumptions needed for process capability, we do not use these C_{pm} values to estimate the percent of results that will be outside medical error limits. However, the C_{pm} ranking suggests that the assays with low C_{pm} values may cause problems and that plots of their data should be examined. We must also remember that C_{pm} values can be arbitrarily improved or made worse by changing the medically acceptable error limits used for each assay.

It was mentioned that sometimes medical acceptability limits may be narrower than quality-control limits. This is telling us that these assays are not capable. It would not make sense to make the quality-control limits narrower for these assays. This is because although more out-of-control alerts would be created, there is nothing to "fix" because the process that produces assay results is working within its inherent capability. However, other choices are to

- Run more replicates of each patient sample, a potential lab user remedy to reduce imprecision;
- Restandardize the assay in some way, a potential lab user or manufacturer remedy to reduce bias;
- Redesign the assay, a potential manufacturer remedy to reduce bias or imprecision.

With respect to any of the above choices, by using Equation 2 one can simulate the effect of running replicates or of improving bias and or imprecision. For

Table 7-1
C_{pm} Results for a Selected Analytes

Analyte	C_{pm}	Total analytical error	Percent error due to bias	Percent error due to imprecision	Percent results outside low limits	Percent results outside high limits
AST	0.281	5.9 U/L	0.7	99.3	10.5	20.0
Cl	0.459	1.81 mmol/L	19.7	80.3	2.9	2.0
Ca	0.620	0.11 mmol/L	4.7	95.3	2.8	2.8
ALT	0.759	5.3 U/L	0.1	99.9	0.9	0.9
Cholesterol	0.882	0.17 mmol/L	4.8	95.2	0	1.9
Alk. Phos.	0.939	5.0 U/L	2.9	97.1	0	0

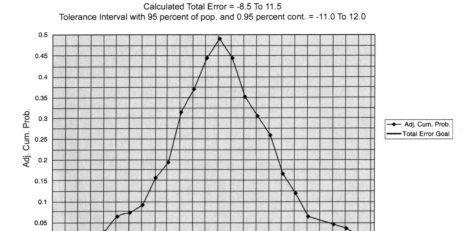

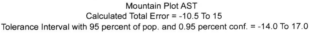

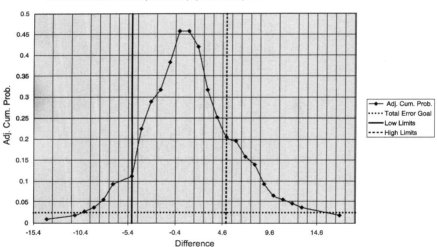

Figure 7-4 Mountain plots for the best (highest) and worst (lowest) C_{pm}s. In the top graph, the low and high limits of -15 and $+15$ are not shown.

assays that have high C_{pm} values, one must always remember that the metric does not capture the random interferences possible in patient samples, so that a high C_{pm} does not guarantee a problem-free assay.

USING COMPLAINTS TO IMPROVE ASSAYS

An important source of data about assay quality is complaint data from people who use the assay results — namely, the clinicians and patients — the latter especially in the case of assays used at home such as glucose monitoring for diabetes. Complaints need not only be performance based. A confusing report format, poor turnaround time, or the failure to run all requested assays represent examples of nonperformance-based complaints. Additionally, every deviation from expected performance may not generate a complaint. Nevalainen and coworkers studied the frequency of laboratory errors and found that certain defects could be as high as 100,000 parts per million (*12*). Reliability tools such as FRACAS, which was presented in Chapter 5, are an effective way to classify complaints so that the limited lab resources can be focused on addressing the most important complaints.

REFERENCES

1. Powers DM. Establishing and maintaining performance claims. Arch Pathol Lab Med 1992;116:718–725.
2. Uniform description of claims for in vitro diagnostic tests. Approved Guideline NCCLS EP11P. NCCLS, 771 E. Lancaster Ave., Villanova, PA, 1996.
3. Pool R. If it ain't broke, fix it. Technol Rev 2001;104:64–69.
4. Pao Y-H. Adaptive pattern recognition and neural networks. Reading, MA: Addison-Wesley.
5. Breiman L, Friedman JH, Olshen RA, and Stone CJ. Classification and regression trees. Belmont, CA: Wadsworth, 1984.
6. Nelson W. Graphical analysis of system repair data. J Qual Technol 1988;20:24–35.
7. Kotz S and Johnson NL. Process capability indices — A review, 1992–2000. J Qual Technol 2002;34:2–19.
8. Boyles RA. The Taguchi capability index. J Qual Technol 1991;23:17–26.
9. Kane VE. Process capability indices. J Qual Technol 1986;18:41–52.
10. Brigden MI and Heathcote JC. Problems in interpreting laboratory tests: What do unexpected results mean? Postgrad Med 2000;107:145–162.
11. Lawton WH, Sylvester EA, and Young-Ferraro BJ. Statistical comparison of multiple analytic procedures: Application to clinical chemistry. Technometrics 1979;21:397–409.
12. Nevalainen D, Berte L, Kraft C, Leigh E, Picaso L, and Morgan T. Evaluating laboratory performance on quality indicators with the six sigma scale. Arch Pathol Lab Med 2000;124:516–519.

Index